THE BOTTOMLESS WELL

Also by Peter W. Huber

Hard Green: Saving the Environment from the Environmentalists

Orwell's Revenge: The 1984 Palimpsest

Galileo's Revenge: Junk Science in the Courtroom

Liability: The Legal Revolution and its Consequences

THE BOTTOMLESS WELL

The Twilight of Fuel, the Virtue of Waste,
and Why We Will Never
Run Out of Energy

PETER W. HUBER
AND MARK P. MILLS

BASIC
BOOKS

A Member of the Perseus Books Group
New York

Published by Basic Books,
A Member of the Perseus Books Group

Books published by Basic Books are available at special discounts for bulk purchases in the United States by corporations, institutions, and other organizations. For more information, please contact the Special Markets Department at the Perseus Books Group, 11 Cambridge Center, Cambridge, MA 02142, or special.markets@perseusbooks.com.

A Note on Graphics
Electronic versions of the charts and graphs, as well as additional related graphics, can be accessed at www.digitalpowergroup.com.

Cataloging-in-publication data for this book is available from the Library of Congress.
ISBN 0-465-03116-1
05 06 07 08 / 10 9 8 7 6 5 4 3 2

For Sophie, Mike, and Steve
P.W.H.

For Tony, Philip, Portia, Brendan, and Donnamarie
M.P.M.

By heavens, methinks it were an easy leap,
To pluck bright honor from the pale-fac'd moon,
Or dive into the bottom of the deep,
Where fathom-line could never touch the ground,
And pluck up drowned honor by the locks.

CONTENTS

FIGURES

PREFACE

WHAT LIES AT THE BOTTOM of the bottomless well isn't oil, it's logic. Fuels recede, demand grows, efficiency makes things worse, but logic ascends, and with the rise of logic we attain the impossible—infinite energy, perpetual motion, and the triumph of power. It will all run out but we will always find more. Some say this is not good for the planet, but that's how it works, regardless. What we will forever seek, and forever find, is not energy but the logic of power.

Many people don't believe this is true, and would be unhappy if it were. "Giving society cheap abundant energy at this point," Paul Ehrlich declared three decades ago, "would be equivalent to giving an idiot child a machine gun."[1] Most right-thinking pundits have since come around to that same view, though few dare put it quite so truculently. The combustion engines that provide our transportation and electricity pollute the air and warm the planet. Protecting our oil supply lines entangles us with feudal theocracies, their bellicose neighbors, and the fanatical sects they spawn. And in any event, our appetite for energy is simply excessive. America consumes 25 percent of the world's natural gas, 23 percent of its hard coal, 25 percent of its crude petroleum, 43 percent of its motor gasoline, and 26 percent of its electricity. We drive more cars many more miles than any other nation on earth. With energy, we would be better off ourselves, and so would the rest of the world, if we simply sought, found, and consumed less.

There are two main schools of thought as to how we should get to less. Cornucopians maintain that through improved efficiency we can have it all—less energy but more light, refrigerated food, warm homes, and safe miles on the highway. By contrast, Lethargists—those in the Ehrlich camp—harbor no such comforting illusions. They know that less really is less, that most people will take more if they can, and that only taxes and regulations will curb the idiot child's appetite.

Both camps are wrong. The Cornucopians are not merely wrong, they are wrong in a spectacularly self-defeating way—energy efficiency leads to more consumption, not less, and if the U.S. government didn't fund it, the Saudis and the big oil companies would. We ourselves will cheerfully join in this camp's celebration of efficiency; we just won't assert that efficiency curbs demand—because it doesn't. It has quite the opposite effect. The Lethargists are wrong too, but more modestly so. More energy consumption isn't worse, it's better. The idiot children are right.

THE DARK 1970s AND THEIR AFTERMATH

In the now standard histories, the beginning of the end of the age of oil arrived on October 19, 1973, when King Faisal ordered a 25 percent reduction in Saudi Arabia's oil shipments to the United States, launching the Arab oil embargo. Supplies were destined to tighten, and prices to rise, from there on out. It would take some time, of course, to lower the curtain. But oil was finished.

The second great energy shock came six years later, on March 28, 1979, with the meltdown of the uranium core of the nuclear power plant at Three Mile Island (TMI) in Pennsylvania. This, all discerning pundits agreed, marked the end of civilian nuclear power in the United States. The Chernobyl accident seven years later just added an extra nail to the nuclear coffin. It didn't matter that the TMI containment vessel had done its job and prevented any significant release of radioactivity, or that Soviet reactors operated within a system that couldn't build a safe toaster oven. Uranium was finished, too.

The nation's first secretary of energy summed things up five months after TMI. "The energy future is bleak," James R. Schlesinger declared, "and is likely to grow bleaker in the decade ahead. We must rapidly adjust our economics to a condition of chronic stringency in traditional energy supplies."[2] Fortunately, the United States could manage on less—much less. Smaller, more fuel-efficient cars were gaining favor, and rising gas prices were curbing demand. And the United States certainly didn't need any more gargantuan electric power plants—efficiency, and the development of renewable sources of power, would suffice.

Another option was to burn an additional 400 million tons of coal a year, which is what we are in fact doing today, over thirty years later.[3] Appliances, air conditioners, refrigerators, and light bulbs grew 30 to 50 percent more efficient in the interim, but all that saving notwithstanding, we still managed to almost double our total consumption of electricity during the same period.[4] Over the same thirty years, the world's oil fields boosted aggregate production by 2.5 billion barrels a year.[5] Oil prices went up and down only modestly in those ensuing years, never approaching the $200/barrel forecast for today by experts in 1980.* And Americans today are burning more of it than ever before.

Most of the new demand for oil was met with imports, but by no means all. U.S. fields, the oldest in the world with many predating World War I, had been scheduled to run dry by the 1990s—only about 30 billion barrels of "proven reserves" remained in 1979, after a century in which about 160 billion barrels (cumulatively) had been pumped out of those same wells. Nevertheless, in the quarter century since 1979, U.S. wells alone yielded another 67 billion barrels. The big oil fields of Oklahoma had been discovered in 1859; the reserves in those fields were

---

*"Energy: A Special Report in the Public Interest," *National Geographic*, February 1981, p. 2—"Conservative estimates project a price of $80 a barrel, even if peace is restored to the Persian Gulf and an uncertain stability maintained." In inflation-adjusted terms, this "conservative" forecast was for $200/ barrel in 2003.

FIGURE P.1   Growth in Total U.S. Fuel Consumption—Post-TMI

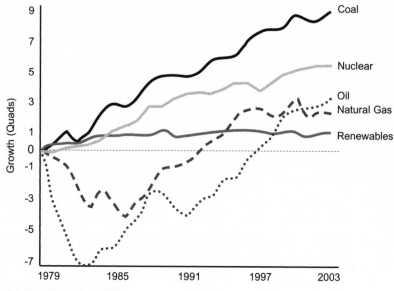

Source: EIA, *Annual Energy Review 2003.*

After the 1979 accident at the Three Mile Island nuclear power plant, many energy pundits concluded that efficiency would curb future demand, and renewable sources would accommodate any future growth. Efficiency did improve dramatically, but demand far outpaced new supplies of renewable fuels. The U.S. now burns an additional 400 million tons of coal every year.

assessed at 125 million barrels in 1969. Yet in the next quarter century they yielded 4.5 billion additional barrels.[6]

As for electricity, coal took care of half of the new demand during that same period, and thus continued to supply over half of our wired power. Electricity generated with natural gas, the fossil fuel grudgingly favored by most environmentalists, dropped sharply for a time but then rose again; it is now back at the 18 percent share it had three decades ago. Astonishingly, it was uranium that advanced the most, pushing its share up from 11 to 20 percent of the electric power generated in the United States. There were seventy-one civilian power plants running in 1980; no new nuclear plants were commissioned after TMI, but others were already under construction. The nuclear count peaked at 112 in 1990. TMI impelled plant operators to develop systematic procedures for sharing infor-

FIGURE P.2 Growth in Fuels Used to Generate Electricity—Post-TMI

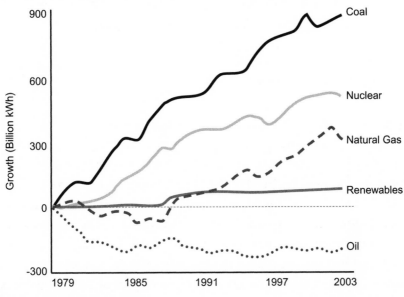

Source: EIA, *Annual Energy Review 2003.*

Coal-fired plants accommodated half of the growth in demand for electricity in the 25 years after Three Mile Island, and continue to satisfy half of the total demand. The accident notwithstanding, nuclear output rose steadily, too. By comparison, gas-fired power has lagged; by and large, gas has merely displaced oil.

mation and expertise, and plants that used to run seven months per year now run almost eleven. Uranium thus displaced about 8 percentage points of oil and 5 points of hydroelectric in the expanding electricity market.

Renewable fuels, by contrast, made no visible dent in energy supplies during this period. About a half billion kilowatt-hours (kWh) of electricity were produced from solar power in 2002, or about 0.013 percent of the U.S. total.[7] Wind power contributed another 0.27 percent. There were subsidies, tax breaks, and government-funded research, but most of the private capital pursued conventional fuels. Fossil and nuclear fuels still completely dominate energy supply in the United States, just as they do in all other industrialized economies.

And what about efficiency? It increased too throughout this period— very rapidly, in fact. Car engines, light bulbs, refrigerator motors—without exception, they all contrived to do much more, with much less. The Cornucopians delivered on every promise but the last. Tremendous improvements in efficiency culminated in more demand, not less.

LETHARGY IN PURSUIT

How about the Lethargists? To be fair, they did at least grasp what would work and what wouldn't. The way to get people to use less energy was to mandate that directly. While it lasted, the national 55 mph speed limit slowed people down and thus limited how far we drove and (indirectly) what we opted to drive. For a while, the Lethargists also made political headway with "fuel-economy" standards for cars and the equivalent for home appliances. Slower trips, dimmer bulbs, smaller refrigerators, and such aren't more efficient; they're slower, smaller, darker—they nudge us toward a less frenetic, peripatetic, and physically expansive way of life. Perhaps this is a good thing. But it is not more efficient, it is more sedentary, calm, and quiet—in short, more lethargic.

It didn't work—at least not on the highway. No one honored the 55 mph speed limit, or the small-car mandates implicit in the fuel-economy standards, which drivers evaded by buying car-like trucks. Ordinary citizens had no direct control over the construction of new power plants, however—this arena is controlled by regulators and those who influence them, and here, the advocates of less did have a real impact.

The advocates succeeded, politically, by making confident projections about the future that we now know were altogether wrong. "Only minor increases in electricity consumption [will] occur" in the future, the Union of Concerned Scientists confidently assured us in 1980.[8] "'Electricity-specific' needs are already met by present capacity with a good deal left over," the ever-quotable Amory Lovins declared in the early 1980s.[9] "The long-run supply curve for electricity is as flat as the Kansas horizon."[10] Pronouncements like these persuaded some regulators, most notably in California, that the nation had built the last big power plant it would ever need. With too many power plants chasing too little demand, prices were bound to hold steady, or fall, even if no new plants were built. The regulated retail price of electricity could be kept very low—this was politically essential—while the construction of new coal furnaces, uranium reactors, and transmission lines could be delayed indefinitely.

But, in fact, electricity consumption would almost double in the two decades after no-more-growth predictions reached their climax. Rising demand collided with flat supply, most dramatically in California in 2000, when prices surged and blackouts rolled across the state. Honest Lethargists could have predicted as much, and perhaps even wished for it. They understood all along that it isn't rising efficiency that curbs demand, still less regulations that keep prices *down*—what curbs demand is whatever pushes prices *up*.

What Lethargists have favored all along are energy taxes.* A tax on energy plainly does promote its antithesis—lethargy. For better or worse, the United States today imposes substantial taxes on fuels and electricity. Taxes increase the average cost of gasoline by 40 percent, and of electricity by 20 to 80 percent, depending on where you live. The first comprehensive U.S. tax on fossil fuel consumption was finally proposed by the Clinton administration in 1993—25.7 cents per million British Thermal Units (Btus), with a 34.2 percent surtax added to oil, gasoline, and diesel fuel, though not to coal, natural gas, hydroelectric power, or nuclear power. Windmills and solar cells were to be exempt from both taxes. These proposals would certainly have curbed energy consumption. They were resoundingly rejected, however, by both houses of Congress, both controlled at the time by Democrats.

European Lethargists have been much more successful—they have pushed energy taxes to the point where their citizens pay roughly twice as much for gasoline and electricity as Americans do. Some Europeans would gladly push them higher still. Germany's Green Party has advanced

---

* "Since energy use, especially use of energy derived from fossil fuels, is central to virtually all of humanity's assaults on its own life-support systems, more general taxes on it would be appropriate." Paul Ehrlich and Anne Ehrlich, *One with Nineveh: Politics, Consumption, and the Human Future* (Island Press, 2004), p. 231.

proposals to triple gasoline prices over the next decade, and to jack up aviation fuel prices apace, so that Germans will fly on holiday no more than once every five years.[11]

Whatever else it may be, this is Lethargy policy at its most candid and straightforward. The problem is defined forthrightly—the problem is energy itself, cheap, abundant energy, efficiently extracted and efficiently used. We humans have too much energy—too much power over creation. Our ever-rising ability to do more things faster, and impose more order of our own choosing anywhere we like, is bad for the rest of the planet, and thus bad in the long run for us too. The only solution is to make energy expensive and scarce.

Western Europe has done much to implement national Lethargy policies; in America, however, we still pursue energy. And because we use the most, we are the most productive and the most powerful. As a result of which, we can, and do, use still more. Perhaps some Lethargists take bitter comfort in the thought that it can't last, the fuels are running out, and the faster we extract and burn them, the sooner the inevitable end must come. But it won't. Humanity is destined to find and consume more energy, and still more, forever.

SEVEN HERESIES

It has not escaped our notice that this proposition, like most of the others we advance, is flatly contradicted by an enormous literature on energy generated since the two great "energy crises" of 1973 and 1979.[12] Most of these books start out with counts of oil wells, coal beds, and gas reserves. Then they observe that conventional supplies are finite and will surely be exhausted sooner or later. They lament our ever-rising demand for energy, expound on how demand could be curbed through readily achievable improvements in efficiency, and demonstrate how much the environment suffers from our failure to curb it. They take it that the nature of "energy" is well understood. Any well-governed society can only want so much of it. One form of energy is much like another, except that some forms are exhaustible and dirty while others are renewable and clean. Energy technology is important, but only insofar as it can raise efficiency, lower demand, reduce pollution, and hasten the transition from old fuels to new.

They are all wrong, except where they aren't even good enough to be wrong, which many of them aren't, much of the time. The best that can be said in their defense is that it is easy to be wrong when writing about energy. "Energy is a very subtle concept," physicist Richard Feynman once observed.[13] "It is very, very difficult to get right. What I mean by that is it is not easy to understand energy well enough to use it right, so that you can deduce something correctly, using the energy idea." Famously plain-spoken though he was, Feynman could have been even more blunt. Most of what most people think they know about energy is so very wrong that their convictions, heartfelt though they may be, lie beyond logical contradiction or refutation.

What most of us think about energy *supply* is wrong. Energy supplies are unlimited; it is *energetic order* that's scarce, and the order in energy that's expensive. Energy supplies are determined mainly by how cleverly we're able to impose logic and order on the mountains and catacombs of energy that surround and envelop us. Supplies do not ultimately depend on the addition of reserves, the development of new fuels, or the husbanding of known resources. Energy begets more energy; tomorrow's supply is determined by today's consumption. The more energy we seize and use, the more adept we become at finding and seizing still more.

What most of us think about energy *demand* is even more wrong. Our main use of energy isn't lighting, locomotion, or cooling; what we use energy for, mainly, is to extract, refine, process, and purify energy itself. And the more efficient we become at refining energy in this way, the more we want to use the final product. Thus, more efficient engines, motors, lights, and cars lead to more energy consumption, not less. Finer, more delicate machines and tasks consume more energy, not less. The transportation of perfectly weightless bits—which are themselves highly ordered packets of energy—accounts for an already significant and rapidly growing fraction of our energy consumption. And however much energy we consume, we will always want more. Demand for energy is as insatiable as demand for information, time, order, and life itself.

Finally, what most of us think we know about the machines that use and transform energy—the *engineering* of energy—is wrong too. Since the dawn of the industrial revolution, the machines have been getting more efficient—and in the aggregate, they have been burning more fuel, too. There is no end in sight to the seeming paradox of rising efficiency and rising consumption; to the contrary, both will rise more in the next few decades than they did in the two centuries since James Watt perfected his steam engine.

These are the seven great energy heresies we propound in this book:

1. *The cost of energy as we use it has less and less to do with the cost of fuel.* Increasingly, it depends instead on the cost of the hardware we use to refine and process the fuel. Thus, we are now witnessing the twilight of fuel.
2. *"Waste" is virtuous.* We use up most of our energy refining energy itself, and dumping waste energy in the process. The more such wasteful refining we do, the better things get all around. All this waste lets us do more life-affirming things better, more cleanly, and more safely.
3. *The more efficient our technology, the more energy we consume.* More efficient technology lets more people do more, and do it faster—and more/more/faster invariably swamps all the efficiency gains. New uses for more efficient technologies multiply faster than the old ones get improved. To curb energy consumption, you have to lower efficiency, not raise it.
4. *The competitive advantage in manufacturing is now swinging decisively back toward the United States.* Steam engines launched the first industrial revolution in 1774; internal combustion engines and electric generators kicked off the second in 1876 and 1882. The third, set in motion by two American inventors in 1982, is now propelling the productivity of American labor far out ahead of the competition in Europe and Asia.
5. *Human demand for energy is insatiable.* We will never stop craving more, nor should we ever wish to. Life is energy in unceasing

pursuit of order, and in tireless battle against the forces of dispersion and decay.

6. *The raw fuels are not running out.* The faster we extract and burn them, the faster we find still more. Whatever it is that we so restlessly seek—and it isn't in fact "energy"—we will never run out. Energy supplies are infinite.

7. *America's relentless pursuit of high-grade energy does not add chaos to the global environment, it restores order.* If energy policies similar to ours can be implemented worldwide, our grandchildren will inhabit a planet with less pollution, a more stable biosphere, and better-balanced carbon books than at any time since the rise of agriculture some five thousand years ago.

THE LOGIC OF POWER

"Energy" appears in the subtitle of this book because that's how the issues we discuss are invariably framed. But in the strict, technical sense of the word, "energy" is completely irrelevant. This book is a chronicle of humanity's struggle against the second law of thermodynamics, not in theory but in the real world, where engineers build practical engines that turn shafts, drive generators, propel cars, run microprocessors, replicate DNA, power heart defibrillators, and project beams of light, radio waves, and X-rays—and yes, of course, engines that also extract the raw fuels that fire the engines themselves. It is a story of ingenious valves and gates that flip open and closed, with just the right timing, to push energy up the thermodynamic hill, to structure our environment, and to add order to our lives.

The book sets out a vision, as well, of the dramatic changes that lie immediately ahead. Energy technology is now poised to evolve faster than at any time before in human history—faster than in 1765, when James Watt invented his steam engine; faster than in 1876, when Nikolaus Otto invented the internal combustion engine; faster than in 1879, when Thomas Edison patented his light bulb. The power of the new millennium is centered on semiconductors: the same materials that made possible digital information have emerged as the enabling materials of digital power. The new technologies of power exploit altogether new physical phenomena to process fantastically concentrated streams of electrons and photons, millions of times faster and far more efficiently than the old technologies they are rapidly displacing.

Over the broad arch of human history, from the nomadic hunter-gatherer to Rome to modern America, the rise of population, life expectancy, great cities, military might, and scientific knowledge has been propelled by rising energy consumption. It is by mastering power itself—the capture and release of energy—that societies master everything else. We rank civilizations accordingly: agricultural societies above nomadic ones, and fossil fuel societies above those that live off the surface of the land. Should they ever become economical, wind, solar, and other green energy technologies will increase our rate of energy capture still more. If they don't, ascendant societies won't adopt them. They will instead favor other technologies that do.

Over the long term, societies that expand and improve their energy supplies overwhelm those that don't. The paramount objective of U.S. energy policy should be to promote abundant supplies of cheap energy and to facilitate their distribution and consumption. Civilization, like life, is a Sisyphean flight from chaos. The chaos will prevail in the end, but it is our mission to postpone that day for as long as we can and to push things in the opposite direction with all the ingenuity and determination we can muster. Energy isn't the problem. Energy is the solution.

ACKNOWLEDGMENTS

THIS BOOK WAS WRITTEN, in part, under the auspices of the Manhattan Institute for Policy Research. We are grateful to the institute's president, Larry Mone, for his patient and unstinting support. We are equally indebted to the many talented engineers, scientists, and business executives, too numerous to list, who told us about the technologies and trends we first explored in writing the *Digital Power Report*, and to George Gilder, who launched and published that technology newsletter. Chuck Davidson and Joe Jacobs of Wexford Capital opened many additional doors when they invited us to help form Digital Power Capital. We received excellent and tireless assistance with research and graphics from Heidi Beauregard, Eileen Oh, Mary Catherine Martin, Jack Levner, and Mark Shaffer. Marilyn Williams supplied essential administrative and logistical oversight.

THE TWILIGHT OF FUEL
AND THE ASCENT OF POWER

I sell here, Sir, what all the world desires to have—POWER.
—MATTHEW BOULTON (1776)[1]

WHAT MOST FRUSTRATES those who feel passionate about energy is that most Americans don't. While we sometimes feel testy about prices at the pump or on a utility bill, few of us get much exercised about the spot price of coal or crude. The reason is simple. Unlikely though it may sound, the cost of energy as we use it today has less and less to do with the cost of the raw fuel that still occupies center stage in the discussion of "energy" policy.

This isn't because free solar and wind power have taken over—they obviously haven't. It's because most of the cost of energy in the form we favor today lies in the processing, the purification, and the conversion. As in a fancy restaurant, most of what we pay for is the fine linen, the service, and the chef's artistry, not the raw calories. With the cost of the power-conversion hardware constantly rising relative to the cost of the fuel, the cost of fuel matters less and less, even as we use more of it. Hence, the "twilight of fuel."

On the official accounts, the United States today consumes about

100 Quads—100 quadrillion British Thermal Units (Btus)—per year of raw thermal energy—and that's leaving out all agricultural sources of energy, the carbohydrates that aren't (yet) hydrocarbons. Roughly speaking, the equivalent of fifteen large, powerful horses labor at peak capacity every second of the day and night for every U.S. citizen. Our appetite for fuel is gargantuan, and it grows geometrically—7 Quads of primary fuel in 1910, 35 in 1950, 100 today, and still rising inexorably year by year.

Yet, year by year, the cost of all those Quads grows less important. The inferno of raw heat that they represent is, paradoxically, slipping into the shadows of our modern economy.

THE BOTTOMLESS WELL

Though he was prepared to go quite a bit deeper when he turned on his steam-powered drill in Crawford County, Pennsylvania, in 1859, Colonel Edwin Drake struck oil at 69 feet. The first "deep water" oil wells stood in 100 feet of water in 1954. Today, they reach through 10,000 feet of water, 20,000 feet of vertical rock, and another 30,000 feet of horizontal rock.

Yet over the long term, the price of oil has held remarkably steady. Ten-mile oil costs less than 69-feet oil did, and about the same as one-mile oil did two decades ago. Production costs in the hostile waters of the Statfjord oil fields of the North Sea are not very different from costs at the historic Spindletop fields of southeast Texas a century ago. There have been price spikes and sags, but they have been tied to political and regulatory instabilities, not discovery and extraction costs. This record is all the more remarkable when one considers that the amount of oil extracted has risen year after year. Cumulative production from U.S. wells alone has surpassed a hundred billion barrels. The historical trends defy all intuition.

It is easy enough to thank human ingenuity for the relatively steady price of a finite and dwindling resource and leave it at that. But there is a second part to this story: it is energy itself that begets more energy. Electrically powered robots pursue new supplies of oil at the bottom of

FIGURE 1.1   Maximum Distance to Oil versus Average Price

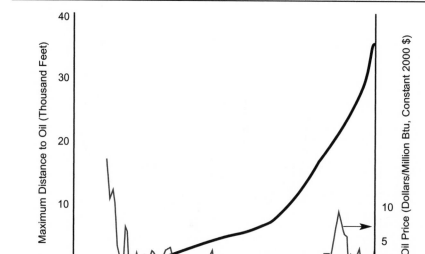

$5/million Btu equivalent to $29/barrel.

Source: WTRG Economics; EIA, *Annual Energy Review 2003*; ExxonMobil; J. Ray McDermott Inc.

Over the long term, the price of oil has held remarkably steady, even as the distance from well-head to the oil has increased from hundreds of feet to miles. Today's production costs in the deep waters of the North Sea are not very different from costs in southeast Texas a century ago.

the ocean. Electricity purifies and dopes the silicon that becomes the photovoltaic cell that generates more electricity. Lasers enrich uranium that generates more electricity that powers more lasers. Power pursues the energy that produces the power.

"Energy supply" is determined not by "what's out there" but by how good we are at finding and extracting it. What is scarce is not raw energy but the drive and the logic that is able to locate, purify, and channel it to our own ends—the creation of still more logic paramount among them. For the first two centuries of industrial history, the powered technologies used to find and extract fuels improved faster than the horizon of supply receded. Hence our blue-whale energy economy. End users consume increasingly compact and intense forms of high-grade power, relying on suppliers to pursue and capture increasingly distant, dispersed, and dilute sources of raw fuel. The gap is forever widening, as the history of oil extraction reveals, but that doesn't stop us—the more energy

FIGURE 1.2   U.S. Oil Prices and Cumulative Production

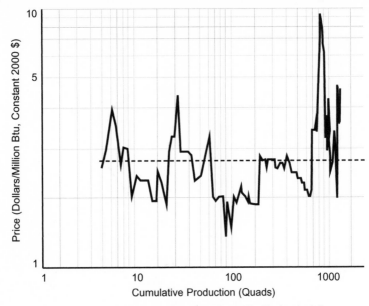

Cumulative U.S. oil production from 1896–2003; 1 Quad=172 million barrels of oil.

Source: EIA, *Annual Energy Review 2003*; American Petroleum Institute; John Fisher, *Energy Crises in Perspective* (Wiley, 1974).

Cumulative production from U.S. oil wells alone approaches two hundred billion barrels. Price spikes and sags have been driven by political and regulatory instabilities worldwide, not by changing discovery and extraction costs.

we consume, the more we capture. It's a chain reaction, and it spirals up, not down. It is, if you will, a perpetual motion machine.

The machine is running faster today than ever before, but it has been running for quite some time. Four billion years ago, life on Earth captured no solar energy at all, because there was no life. Life then got a foothold, and the capture and consumption of energy in the biosphere has been rising ever since. The thicker life grew on the surface of the planet, the more energy the biosphere managed to capture. And it used all that energy to create more life. Along the way it deposited huge amounts of biological debris underground. A new form of life then emerged, a scavenger capable of feeding not just on fresh carrion but on the debris itself. James Watt invented a machine to dig up the debris more efficiently—his coal-fired steam engine was designed, initially, to pump out the water that kept flooding the coal mines. In exactly the same manner, though on a far tinier scale, Enrico Fermi built the first fission reactor by using one neutron emitted by a uranium atom to kick out two neutrons from other uranium atoms nearby.

None of these processes produces "perpetual motion" in the strict thermodynamic sense, of course—all of them just improve on the process of seizing energy from somewhere else. Living green plants still capture today's solar energy about six times faster than we humans are able to dig up yesterday's solar energy preserved in fossil fuels, but we'll overtake the rest of nature in due course. Perhaps someday we'll get to the point where we, too, can capture our energy directly from the sun. There's plenty of sunlight to spare—green plants currently capture only about one three-thousandths of the golden cascade of solar power that reaches the Earth's surface.[2]

But whether we catch our solar energy live, dig it up in fossilized form, or mine uranium instead is really just a detail. The one certainty is that we will extract more energy from our environment, not less. Everything we think we know about "running out of energy" isn't just muddled and wrong; it's the exact opposite of the truth. The more energy we capture and put to use, the more readily we will capture still more.

THE ENERGIZED ENVIRONMENT

Energy isn't scarce, and it doesn't cost anything. The world's first uranium reactor was assembled some 1.8 billion years ago, near what is now the Oklo River in Gabon, Africa. At the time, the Earth's crust contained much more of the fissionable uranium isotope (U-235) that fuels reactors today. Chemically concentrated, by slow bacterial action, perhaps, then flooded by fresh water from the river, natural uranium oxide supplied the right mix of fissionable fuel and neutron-slowing water to start a chain reaction. Half a dozen separate reactors formed spontaneously at Oklo; each site generated 5 to 10 kilowatts (kW) of heat for the next fifty to one hundred thousand years. Several of the reactors transmuted more than half of the original U-235 into the telltale fission by-products that are still there today.

Some time later—in 1933, to be precise—Leo Szilard conceived of the possibility of doing much the same thing. Nine years later, Enrico Fermi built a second Oklo, this one on a squash court under Stagg Field Stadium at the University of Chicago. In 1956, radiochemist Paul K. Kuroda suggested the possibility of such reactors forming spontaneously in the Earth's crust; French chemists, geologists, and physicists would later set out in search of such a site; and the astonishing discovery at the original Oklo was documented in *Scientific American* in 1976.[3] The total heat released over the course of 50 or 100 millennia at Oklo itself could have powered a town of 250,000 for about a year. Natural radioactivity in the Earth as a whole produces about 1,000 Quads of thermal energy per year, some small part of which we see at the surface in the form of volcanoes and hot springs.

Far more—an estimated 5 million Quads per year—of solar energy reaches the surface from above[4]—roughly ten thousand times as much as humanity consumes in the form of fossil fuels, crops, and wood.* Green plants seize and temporarily store a tiny fraction of it. During the night, the dark side of the Earth radiates all the rest, along with the geothermal heat, back out into the black depths of the cosmos.

But if energy is so abundant, why does it cost so much at the gas pump? If bacteria can build nuclear reactors, why do electric utilities send us hefty bills every month? The short answer at the pump is that the price of gasoline has nothing at all to do with the abundance of raw energy, and surprisingly little to do with the abundance of crude oil in the Earth's crust. As for utility bills, the Oklo microbes may have concentrated the fuel a bit to liberate heat, but their heat didn't turn a shaft to spin a generator to move electrons. The bacteria beat Enrico Fermi to the punch, but to power the lights in San Jose you need a James Watt and a Thomas Edison too, and that gets expensive.

---

*There are a half dozen different units for measuring energy—joules, Btus, calories, watt-hours, horsepower-hours, foot-pounds—each used by different conventions in different fields. We use Btus or watt-hours (3,412 Btu = 1,000 Wh or one kWh) throughout this book.

All we learn from Oklo is that there are unimaginably large amounts of energy out there, stored in so many places and flowing so ubiquitously that even bacteria and plankton can stumble upon it, unleash it, and thrive on it. But we didn't have to travel to Oklo to prove that. Life has thrived so abundantly on the vast supplies of energy so readily at hand that life now colors much of our planet green.

Pyramids of Quads

Humans are opportunistic scavengers of energy, like all other forms of life. Before we burned fossils, we burned life itself—trees, whale oil, and grass, by way of horses, for horsepower. As late as 1910, some 27 percent of all U.S. farmland was still devoted to feeding horses used for transportation.[5] Feeding that organic transportation system required twice as much land as we use today for all our roads and highways, oil pipelines, refineries, and wells.[6] Much of the rest of the world still runs on a carbohydrate energy economy—the power of the horse, oxen, camel, or donkey, and the manual labor of peasant, indentured servant, or slave.

Then we learned how to burn dead plants too—biomass that has been compressed and refined for hundreds of millions of years in geophysical refineries. And then, the elemental constituents of the Earth itself. A Quad's worth of wood is a huge forest—beautiful to behold, but bulky and heavy. Pound for pound, coal stores about twice as much heat. Oil beats coal by about twice as much again. And a gram of U-235 is worth about four tons of coal. The proponents of solar, wind, biomass are pushing against a powerful historical trend. Left to its own devices, the market has not pursued thin, low-energy-density fuels, however cheap—it has paid steep premiums for fuels that pack more energy into less weight and space.

The energy-transforming machines that burn these primary fuels have evolved in exactly the same way—toward more power in less space, or higher *power density*. Oil derricks, car engines, and microprocessors all run much faster than they used to and handle more power in less space. An old-fashioned steam locomotive requires a large, heavy coal furnace,

FIGURE 1.3   Energy Density of Primary Fuels
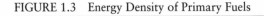

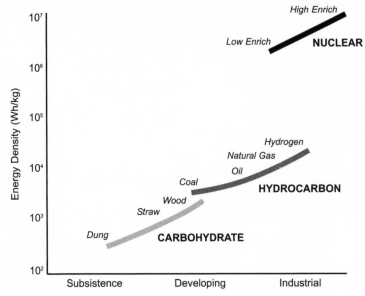

Source: Smil, *Energies* (MIT, 1999); B. Spletzer, "Power Systems Comparisons, Intelligent Systems and Robotics Center," Sandia National Laboratories, Sept. 1999.

A Quad's worth of wood is a huge forest. Pound for pound, coal supplies about twice as much heat. Oil is twice as good as coal. And a gram of uranium-235 is worth about four tons of coal. Historically, we have always pursued fuels that pack more energy in less space.

and it reciprocates slowly; the gas turbine under the wing of a jumbo jet burns much better fuel, much faster, to produce far more power in far less space.

We have in fact advanced so far that—in our official statistics, at least—we now maintain that "horsepower" has nothing at all to do with the muscles of human or horse. We casually assume that biking and walking entail no "energy" overhead; we do not pause to consider that a tree might have been cut down to clear space to grow the granola that supplied the calories that the biker now labors to burn off. Our officially reported 100 Quad energy budget simply ignores our own carbohydrate energy economy as too small to matter. When we compare our own energetic excesses to the developing world's supposed frugality, we generally ignore their carbohydrate fuels and the (considerable) environmental impacts those fuels entail.

FIGURE 1.4   Power Density from Horses to Lasers

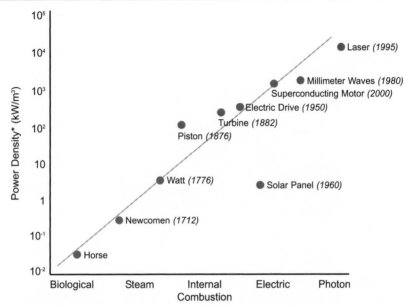

The horses, engines, motors, jets, antennas, and lasing cavities that transform primary fuels into motion and beams have evolved on a trajectory of more power in less space.
*for complete power system

Pontiacs, Power Plants, and Pentiums

So what do we actually do with the 100 Quads we burn every year? Picture a vast parking lot filled with 10,000 Pontiacs. At exactly the same moment 10,000 drivers start their engines, shift into neutral, and floor their accelerators. All 10,000 Pontiac engines rev up to the red line on the tachometer. These Pontiac engines are now generating a total of about 1 gigawatt (GW) of kinetic power. Power plants on wheels burn a lot of fuel—but in percentage terms, somewhat less than we may suppose—about 30 percent of the national intake, or 30 Quads of raw fuel.

Alongside the parking lot stands an electric power plant—a fairly big one, but just one, with a few dozen high-voltage cables—a single transmission "line"—leading out of it. That single line carries about one gigawatt of electric power—the same as the 10,000 drive shafts of the 10,000 red-lining Pontiacs. If we drove our 220 million Pontiac-sized

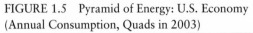

FIGURE 1.5   Pyramid of Energy: U.S. Economy
(Annual Consumption, Quads in 2003)

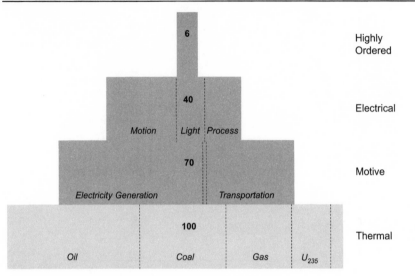

Source: EIA, Annual Energy Review 2003.

Energy consumption is often totaled up in Quads—thermal units of raw heat. But this metric conceals more than it reveals. In the United States, about 6 Quads worth of raw thermal energy go through multiple stages of refinement every year, with 99 percent (or more) of the original thermal energy being discarded along the way, to yield the extremely highly ordered power required to drive such things as radios, microprocessors, lasers, and CAT scanners.

power-plants-on-wheels flat out around the clock, they would burn 5 to 10 times as much fuel as the 5000 (or so) electric plants that generate all our electricity. But the electric plants in fact burn more Quads—about 40 Quads per year, or 40 percent of our total—and produce much more useful power, because they run very efficiently, around the clock, and at close to peak capacity for many more hours of the day.

Power plants and Pontiacs define two of the three basic things we do with our 100 Quads of fuel: we make electricity, we move vehicles, and we produce raw heat. In round numbers, the percentage mix is 40/30/30. From the get-go, then, it's apparent that Quads don't measure the important stuff. Quads measure "thermal units"—raw heat—yet in our economy, scarcely one Quad in three ends up being used as heat itself, in such things as ovens, dryers, and welders, the devices that heat air, water, foods, and chemicals, that dry paints, forge steel, and weld ships.

FIGURE 1.6   Primary Fuel Uses

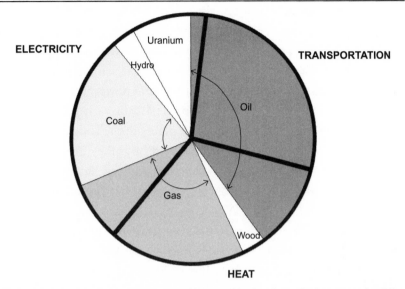

We use primary fuels to make electricity, move vehicles, and produce heat. Coal generates electricity. Oil is the fuel of transportation. Natural gas is used principally for raw heat; it also now generates 18% of our electricity.

All the rest of the Quads are used in combustion engines to spin shafts—raw heat transformed into orderly motion. Somewhat under half of that shaft power is used for transportation; the rest is transformed yet again—it spins the generators that produce electricity. And some share of the electricity—about 6 Quads' worth, if we continue to track everything by the amount of thermal fuel used at the outset—goes through additional stages of refinement, to yield the extremely well-ordered power required to drive such things as radios, microprocessors, lasers, and CAT scanners.

That we use 40 percent of our raw Quads to generate electricity, and only 30 percent for transportation, is revealing too. Cars are everywhere, while electric power plants are all but invisible—do our light bulbs, refrigerator motors, and so forth really burn more fuel, however indirectly, than our SUVs? Yes. We have far fewer power plants than cars, but the power plants are much bigger, and we run them much harder.

Oil is of course the fuel of transportation. Natural gas is used principally to supply raw heat, but it is now also making steady inroads into electric power generation. Electric power plants are fueled mainly by coal (which generates about 55 percent of all our electricity) and uranium (20 percent), in massive 500 to 1,000 megawatt (MW) power plants. Another 18 percent of our electricity is generated by somewhat smaller, 60 to 180 megawatt gas turbines—basically the same engines that hang under the wings of a jumbo jet. Hydroelectric power is a factor too (under 10 percent), but oil is almost completely absent from the electricity slice of the pie chart, at least in power plants in the United States.

These sharp divisions among fuels and how they are used emerged relatively recently. Developing economies still rely largely on growing plants—trees and pasture—for heat and transportation, and don't generate much electricity at all. A century ago, coal was the all-purpose fuel of industrial economies: coal furnaces provided heat, and coal-fired steam engines powered both trains and the early electric power plants. From the 1930s until well into the 1970s, oil fueled not just cars but many electric power plants too. By 2020, as we will argue in later chapters, electricity will almost certainly have emerged as the new cross cutting "fuel" in both stationary and mobile applications.

Today, however, each of the three main sectors of energy consumption uses different fuels to power quite different conversion technologies. External combustors and turbines propel the electrons; internal combustors and piston engines propel the cars; furnaces and open flames supply plain old heat. About 60 percent of the fuel we use is not oil—it's coal, uranium, gas, and hydroelectric, and most of this not-oil fuel is used to generate electricity.* The other 40 percent of the fuel we use is indeed oil, most of which is used by cars. The not-oil fuels go mainly into a few thousand gargantuan power plants. The oil, along with much of the gas, is channeled mainly into hundreds of millions of much smaller car engines, boilers, and furnaces.

And the strangest thing about *that* is that as fuels go, electricity isn't

---

*To put the numbers a bit more precisely, about 40 percent of our fuel is oil; coal and natural gas weigh in at 24 percent each, uranium is 8 percent, and hydroelectric power accounts for most of the remainder. For all the hopeful attention they attract, wind, solar, and other small-scale "renewables" are still scarcely discernable in the overall energy accounts.

cheap at all. Indeed, if price is measured in terms of raw energy, electricity is far more expensive than mechanical or thermal alternatives. Yet for reasons that seem to defy common economic sense, we keep pushing our consumption of energy toward the more expensive alternative.

But they defy common sense only so long as we speak in terms of energy—which is irrelevant—rather than energetic order, a fundamentally different (and much more difficult) concept, which explains everything. In brief, the *order* in the energy is the only part that has any value. The sun provides us with 100 watts (W) of light for free, through a couple of square feet of skylight, at noon on a moderately sunny day. Yet we pay good money for a 100 watt bulb and the electrons to light it. And thousands of times more for a 100 watt laser and its power supply. A photon is a photon, but better-ordered photons packed into less space, and delivered on demand, are worth far more than the diffuse, disordered, episodically available alternative, however "renewable" the sunlight may be.

This, too, is why we abandon wood in favor of oil—because oil packs more energy into less space. And why we refine oil further, to separate the gasoline from the tar, and why we use massive arrays of centrifuges to enrich uranium. Year by year we shift a larger share of our energy consumption from other alternatives to electricity—which is by far the most dense "fuel" that we commonly use and also by far the most expensive. These choices don't add energy, they subtract it—oil refineries consume substantial amounts of power, and electric power plants dump half or more of all the heat from their furnaces into cooling towers and nearby rivers. But we channel more and more of our energy through these systems anyway. The laser used for cataract surgery does nothing but convert energy from one form to another, and makes no sense at all in raw energy terms—far more electrical power is sucked into the back end of the laser than gets pumped out as an intense beam of light at the front. But a laser beam can restore eyesight; electricity can't, nor can heat from the coal boiler still lower down in the energy pyramid.

This is obvious, almost self-evident—and yet none of the conventional metrics of energy and power, the obsolescent accounts of Btus, octane-rated gallons, therms of natural gas, kilowatt-hours (kWh), or calories, captures this, the most fundamental fact of all.

FIGURE 1.7   Pyramid of Spending
($ Billions, Constant 2000 $)

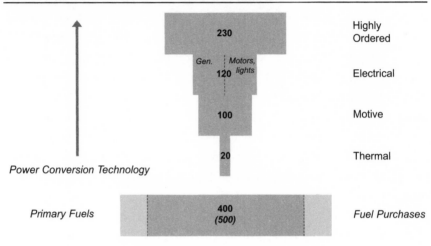

America spends about $400 billion a year on raw fuel ($500 billion with oil priced in the $50/bbl range). But at least $500 billion a year is spent on equipment used in the higher tiers to concentrate and convert energy—generators (gen.), furnaces, car engines, motors, and light bulbs, for example. The largest and fastest-growing segment of the power-conversion economy now comprises such things as power semiconductors, lasers, ultrasound machines, magnetic resonance imagers, and telecommunications equipment that produce highly ordered power.

CAPITAL SPENDING

Electricity is expensive because it takes a huge amount of expensive hardware, and a great deal of fuel, to refine raw fuel into electric power.

This hardware—this enormous energy-converting capital infrastructure —is a relatively recent development. Until the rise of the steam engine, virtually all "energy technology" was directed at extracting Quads from the biosphere—harvesting crops and husbanding animals. Through the nineteenth century and well into the twentieth, most of our energy spending went to capturing energy at the surface of the Earth, or digging it up from beneath the surface—much more was thus spent on getting the raw fuel than on the ovens, engines, and other hardware that converted it to energy of another form.

In the last few decades, however, that picture has been turning upside

FIGURE 1.8   The Price of Power

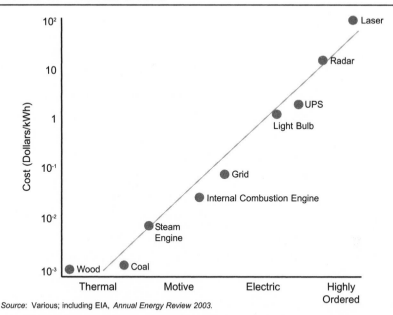

Source: Various; including EIA, *Annual Energy Review 2003*.

The price of power rises sharply with rising power density. Relatively unreliable grid power retails for 10 cents per kWh. The same amount of thermal energy locked up in raw coal costs about 1/3 of a cent. Computer-grade power (UPS) costs $3 or more per kWh.

down—today, we spend almost as much, or more, on the conversion hardware. In recent years, America has been spending about $400 billion a year on raw fuel, oil priced at $50 per barrel going forward would push the total to about $500 billion. But we spend at least $500 billion a year on new capital equipment[7] used in the higher tiers to concentrate and convert energy—furnaces, generators, car engines, motors, light bulbs, lasers, and so forth—that transform heat into motion, motion into electricity, and electricity into light or back into motion.* The largest and fastest-growing segment of capital spending is for a new tier of hardware that produces very-high-quality electricity, sound, microwaves, laser light, X-rays, magnetic pulses, and such—power precise

*In addition, about 10 percent—$50 billion or so a year—of the cost of the raw fuel is attributable to the capital equipment used to extract and refine it—offshore oil platforms, coal excavators, remotely operated undersea robots, oil refineries, and the like.

enough to activate gigahertz-speed (GHz) silicon chips, lasers, ultrasound machines, magnetic resonance imagers, and high-speed wireless telecommunications systems.

The price of power rises sharply with each step up this pyramid. Relatively unreliable grid power retails for 10 cents per kilowatt-hour. The same amount of thermal energy locked up in raw coal costs about one-third of a cent. Computer-grade power, backed up by uninterruptible power sources and many additional layers of power-conditioning electronics, costs $3 or more. And yet the amount of electricity we use rises and rises, with consumption of the very highest grade electricity rising fastest. Today, Americans consume about $230 billion per year of retail electric power—over one-half the total America spends on all the raw fuel that is burned in all power plants, cars, and heating systems combined.[8] And outlays for energy hardware in the upper tiers of this inverted pyramid are rising much faster than spending on raw fuel at the bottom.

THE REAL COST OF POWER

At the end of the day, the consumer doesn't much care about the cost of just the fuel or the hardware—the consumer wants what they deliver together. The more we spend on the hardware, the less we notice the cost of the fuel.

This is obvious enough with cars, even though they remain wholly dependent on the volatile market for oil. The lifetime cost of the car and its engine substantially exceeds the lifetime cost of its fuel. As car manufacturers and their critics both well know, few drivers pay much heed to gas-mileage stickers, and that indifference is easily explained: fuel costs represent under 20 percent of the typical cost of driving. The largest cars guzzle the most gas, but the price of the hardware rises faster than the mileage falls, and gas mileage thus matters the least to those who burn the most fuel in their luxurious sedans and SUVs. The roughly 5 percent gas-guzzler excise tax that the federal government imposes on vehicles that fail to meet the Environmental Protection Agency (EPA) mileage targets hardly changes the equation at all.

The schism between power and fuel is sharper still in the rest of our

FIGURE 1.9   U.S. Energy Production and Cost

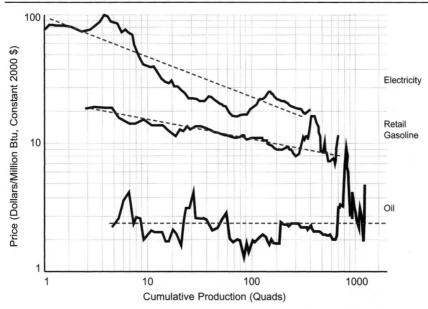

Source: EIA, *Annual Energy Review 2003*; American Petroleum Institute; John Fisher, *Energy Crises in Perspective* (Wiley, 1974).

Fuel prices have fluctuated, but new fuels and improvements in power-conversion hardware have steadily lowered the average retail price of electricity ever since Edison fired up his first generators in 1882. 1 million Btu of electricity is a 60 W light bulb running for 6 months; 1 million Btu of gasoline is about 8.1 gallons, and 1 million Btu of crude oil is about 7.1 gallons.

energy economy, which is much less dependent on oil. The nontransportation sector of our gross domestic product (GDP) (about 90 percent of the whole) already gets over half its energy from electricity. And the price of electricity does not depend strongly on the price of the primary fuels that generate most of it.

Oil is largely out of the electricity side of the picture, because it generates under 5 percent of U.S. electric power. Electricity prices aren't strongly tied to the price of coal or uranium either, which together generate 75 percent, because most of the cost of power lies in the power plant and the distribution system—in capital and logic, not in combustible chemicals or fissile atoms. All in all, raw fuel accounts for over half the delivered cost of electricity generated in gas-fired turbines, about one-third of coal-fired power, a tenth of nuclear electricity, and none of the cost of hydroelectric and solar power.

Over the long term, the hardware has grown steadily more efficient. Thus, even as fuel prices have fluctuated and fuel mixes have shifted, the average retail price of the kilowatt-hour has fallen almost without interruption since Edison fired up his Pearl Street generators in New York in 1882. When electricity prices have gyrated wildly, it has been because of regulatory and political factors, not fuel supplies or technology. The long-term price trends are unambiguous: the price of a kilowatt-hour keeps dropping.

THE TWILIGHT OF FUEL

The economic importance of fuel in the conventional sense of the word —which is to say, stuff that gets very hot—will continue to recede because electricity is the ascendant mover at the front end of our modern economy. Oil still defined the core of our energy economy when President Jimmy Carter exhorted us all to wear sweaters, but electricity defines the epicenter of energy markets today. More than 85 percent of the growth in U.S. energy demand since 1980 has been met by electricity. About 60 percent of our GDP now comes from industries and services that run on electricity—in 1950, the figure was only 20 percent. Some 60 percent of all new capital spending is on information-technology equipment, all of it powered by electricity. All the fastest growth sectors of the economy—information technology and telecom, most notably— depend entirely on electricity.

And the electrification of our energy economy is accelerating.[9] To begin with, electricity is taking over major parts of the thermal sector. A microwave oven can displace much of what a gas stove is used for in a kitchen; lasers, microwaves, magnetic fields, and other forms of high-intensity photon power provide more precise, calibrated heating than conventional ovens in manufacturing and industrial processing of materials. Lasers now move ink in printers, etch silicon and metal, solder optoelectronic chips, cure epoxies, bond polymers, weld, sinter, burn hair, cauterize tissue, and reshape the surface of the eye. Over the next two decades, these trends will move about 15 percent of our entire energy economy from conventional thermal processes to electrically powered ones.

FIGURE 1.10   Fuels for the Economy*

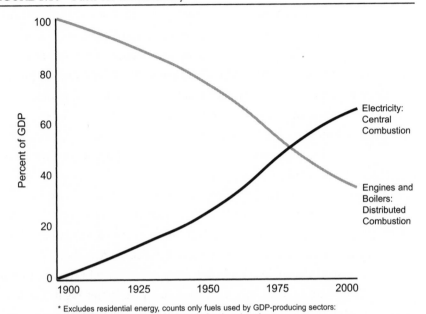

* Excludes residential energy, counts only fuels used by GDP-producing sectors: transportation, industry (incl. mining, agriculture), and services.

Source: EIA, *Annual Energy Review 2003*; Bureau of Economic Analysis; U.S. Census Bureau, *Historical Statistics of the United States Colonial Times to 1970.*

Two-thirds of the U.S. GDP now comes from industries and services fueled by electricity. All the high-growth, information-centered sectors of the digital economy run entirely on electricity.

Electricity is also taking over the power train in transportation—not the engine itself, but the system that moves power throughout the car. Diesel-electric locomotives and many of the monster trucks used in mining have already made the leap to electric drive trains; the oil-fired combustion engine is still there, but now it's just an on-board electric generator that propels nothing but electrons. The transition to the hybrid electric car will be completed over the next two decades as well. During this same period, electric power trains will supersede steel shafts, belts, pulleys, and hydraulic systems in factories.

We still need fuel to get the flow of power started, of course. The laser welder in the factory will still depend on something up the line to power the grid; the hybrid-electric car will still require gasoline, diesel, hydrogen —something—to generate its supply of on-board electricity. The most optimistic green visionaries foresee solar cells and wind turbines generating the electricity of the future, with water, hydrogen, and fuel cells being used in between to get the electrical energy into and out of the fuel tanks of cars—again, we will return to this possibility in chapter 5. Conceivably, as we discuss further in chapter 6, the electric car in the driveway will also be able to power the home alongside, thus bridging the present fuel chasm between electricity and transportation. Power plant, piston engine, coal, uranium, and oil give way to sun, wind, and water. It is a charming vision.[10] And if by chance it comes true, wind and sun are of course free, so the *entire* cost of electricity will end up being determined by the hardware alone.

Practical-minded utilities foresee, instead, an additional two hundred one-gigawatt power plants, still fueled mainly by coal and uranium. Utilities used to tout the cheapness of these fuels too, and with much the same logic in mind. Nuclear power was going to be too cheap to meter, and if the cost of the fuel was all that mattered, that day would now be close at hand—closer, certainly, than too-cheap-to-meter solar. Uranium ore is plentiful and cheap; enrichment is the main challenge in turning it into a fuel and accounts for about one-half of the fuel's cost, and the fuel thus enriched accounts for a much smaller fraction of the cost of nuclear electricity. The fuel that spins the turbines alongside the Hoover Dam is free too; all the cost of hydroelectricity is attributable to the cement and the steel.

For present purposes, it just doesn't matter which of these two visions comes true. Either way, the cost of energy as we actually use it will become further separated from the cost of fuel. Either way, Hoover Dam economics will continue to overtake our electricity supply, with more and more of the cost in the hardware, less and less in the fuel.

And either way, the shift to electricity is happening because the technologies are at hand to make it possible, and because electricity can do more, faster, in much less space—it is by far the fastest and densest form of power that has been tamed for ubiquitous use. Electricity moves at the speed of light; all other forms move at the speed of sound, or slower. The ten thousand Pontiacs in the parking lot require ten thousand steel drive shafts to transmit their power. The single power plant dispatches the same amount of power through a few dozen high-voltage wires.

So the electron advances on every front, while the mechanical and thermal forms of power retreat. They don't disappear—to the contrary, they're required more than ever to generate all the electricity—but we see much less of them close at hand, and we have less and less reason to worry about their cost. The rise of all this complex, energy-conversion hardware fundamentally changes the basic architecture and economics of our energy economy. When half-cent-per-kilowatt-hour coal in an industrial boiler finally gives way to $200-per-kilowatt-hour photons from an electrically powered ytterbium laser, the cost of coal, uranium, sun, and wind hardly matters any more at all.

THE FIRST LAW

And yet . . . it still seems so strange, so unreal, this proposition that the more energy-processing machines we build, and the more Quads of raw energy those machines burn and refine, the less we find ourselves worrying about "energy" as that term is ordinarily used. Common sense, backed by rock-solid axioms of economics, would seem to require just the opposite. What gives?

"Energy," it turns out, is a far more subtle concept than is commonly recognized outside of engineering schools and physics departments. In discourse about public policy, the concept is so poorly defined, and the word is so casually misused, that most of what we think we know about "energy" is wrong, if not incoherent.

In 1964, Richard Feynman, the future Nobel physicist, found himself reviewing school textbooks for California's Curriculum Commission. As he described in two subsequent (and hilarious) accounts, it was a "horrifying" experience.[11]

> There was a book that started out with four pictures: first there was a windup toy; then there was an automobile; then there was a boy riding a bicycle; then there was something else. And underneath each picture it said, "What makes it go?"
>
> I thought, "I know what it is: They're going to talk about mechanics, how the springs work inside the toy; about chemistry, how the engine of the automobile works; and biology, about how the muscles work." . . .

I turned the page. The answer was, for the windup toy, "Energy makes it go." And for the boy on the bicycle, "Energy makes it go." For everything, "Energy makes it go."

Now that doesn't mean anything. Suppose it's "Wakalixes." That's the general principle: "Wakalixes makes it go." There's no knowledge coming in. The child doesn't learn anything; it's just a word!. . .

It's also not even true that "energy makes it go," because if it stops, you could say, "energy makes it stop" just as well. What they're talking about is concentrated energy being transformed into more dilute forms, which is a very subtle aspect of energy. Energy is neither increased nor decreased in these examples; it's just changed from one form to another. And when the things stop, the energy is changed into heat, into general chaos.

Heat is indeed "general chaos," just as Feynman says. Why then do we pay so much for raw fuel, which generates nothing but heat? Why do we measure our entire energy budget in Quads—Quadrillions of British *Thermal* Units—which is to say, units of heat? And why, when there's too much heat energy in our living room, do we rush to buy still more energy from someplace else? About one-half of the electricity we use runs motors, and nearly half of that electromotive power is used to run air conditioners and coolers. The only useful thing these expensive, energy-devouring machines do is help us get rid of energy we don't want. The room is too hot so we burn some coal in the furnace of a power plant a hundred miles away to run the motor that compresses the coolant that removes the solar heat from our living room. When the job is finished, we have dumped energy up the smokestack of the power plant in order to dump energy out the window of our living room.

Why do we ever have to buy energy, or dig it up, more than once? The first law of thermodynamics declares that energy can't be created or destroyed, it just flows from here to there. All the energy that starts as chemical in your gas tank ends up as heat in the atmosphere, or the brake pads, or the car's tires, or the tarmac, or in some small change in the speed at which the Earth rotates. Why then must we keep worrying about energy supplies?

As we will discuss in chapter 3, most of the 100 Quads of energy that we currently consume ends up in the dump, too—not after we're done using them, but before the using even begins, because most of the Quads are used for no purpose other than to purify energy itself. The power-purification overhead is staggeringly high—some 80 to 95 percent of the energy we use never moves a useful payload like the driver in the car, never emerges from a glowing filament as a useful lumen of light, never leaves an antenna as useful electromagnetic waves, never heats food in an oven or cools it in a refrigerator, never makes it to the final point where it actually gets put to human ends.

If this sounds unlikely, it is because the laws of thermodynamics defy all ordinary intuition. The first law holds that energy is always conserved*—the "conservation" of energy is preordained—what we really aspire to do is to conserve energetic order, not energy itself. The gasoline in the tank before the drive is worth much more than the heat in the tires, tarmac, and brake pads at the end of the trip. The heat, in fact, is worth less than nothing; it destroys things and we spend extra to get rid of it.

And that, in an abstract but necessary nutshell, explains the twilight of fuel. Raw fuel set off into the twilight in 1776. When Johnson's biographer, Boswell, visited the Boulton-Watt works that year, Matthew Boulton explained it all, in the famous double entendre quoted at the head of this chapter. Boulton and his partner James Watt were selling "what all the world desires"—not fuel but power, a quite different, more elusive, and much more expensive resource, not raw, stored energy, but well-ordered energy in motion, not heat, but a spinning steel shaft. Fuel has been receding into the shadows ever since, as our appetites have shifted progressively to ever purer, denser, better-ordered forms of power. Year by year, we spend more and more on the process of converting energy to power, and power to still purer power, while the cost of the fuel itself fades into the footnotes of our economic ledgers.

---

*For purists, this law is violated only in nuclear processes at the core of stars and reactors wherein energy indeed is created through the destruction of matter. Nothing else in the energy arena—annual perpetual motion patent filings notwithstanding—and nothing else in life or on Earth violates the first law.

2

VORACIOUS TECHNOLOGIES

The business I am here about has turned out rather successful; That is to say that the fire-engine I have invented is now going, and answers much better than any other that has yet been made & I expect will be very beneficial to me.

—JAMES WATT, LETTER TO HIS FATHER, DECEMBER 11 (1774)[1]

My friend Herschel, calling upon me, brought with him the calculations of the computers, and we commenced the tedious process of verification. After a time many discrepancies occurred, and at one point these discordances were so numerous that I exclaimed, "I wish to God these calculations had been executed by steam!"

—CHARLES BABBAGE (1821)[2]

THE AGE OF FOSSIL FUEL began in the fecund mind of a Scottish engineer in 1774. Until then, the energy economy had been centered on carbohydrates—wood and pasture. A book like this one written in 1774 might well have started out with some Malthusian reflection on why our population would always grow faster than its supplies of energy. In his 1798 *Essay on the Principle of Population*, Thomas Malthus himself focused entirely on agriculture, or what today's energy pundits would call "biomass." But the next century of energy would in fact be defined by the spiraling new demand for coal created by James Watt's new steam

24

engine. An energy book written in 1876 could have been all about coal—overlooking Nikolaus Otto's new invention of the first practical internal combustion engine and the coming century of oil that it fore-shadowed. The pundit of 1879 might perhaps have foreseen that Edison's new light bulb would spur some demand for new-fangled electricity, but could hardly have imagined the air-conditioning loads of the 1950s or the personal computer loads of the 1990s.

As Richard Feynman saw, "energy" is much too subtle a concept to start with in the search for understanding. Framing public policy around such a very difficult abstraction is harder still. It is not possible to think intelligently about supply, demand, or what ought to be, without under-standing the things that use and process energy—the engines, motors, bulbs, and computers—and where those technologies might be headed. But anticipating where technologies are headed is never easy, least of all in times of rapid technological change.

This much we do know: our sharply rising consumption of fossil fuels can be traced back to specific individuals: James Watt, Nikolaus Otto, Rudolf Diesel, and Henry Ford. Likewise our consumption of electricity —to Thomas Edison, George Westinghouse, and Nikola Tesla. It was be-cause of them that U.S. consumption of Quads (raw energy) rose ten-fold in the last century, and because of them that consumption of electricity rose thirty-fold during the same period. With the technologies now at hand, it is safe to predict that today's totals—100 Quads and almost 4 trillion kilowatt-hours (kWh) per year—will reach 130 Quads and 5 trillion kilowatt-hours by 2020. Or more.

THE "FIRE-ENGINE"

There were steam engines long before James Watt arrived on the scene. What made Watt's "fire-engine" so much better wasn't fuel or steel but logic, embodied in what Watt called an "engine regulator." The regulator opened and closed valves in synchrony with the up-and-down motion of the piston, by way of a system of gears, connected to an axle, connected to the piston itself. The regulator thus controlled the flow of

FIGURE 2.1   U.S. Energy and Electricity Consumption

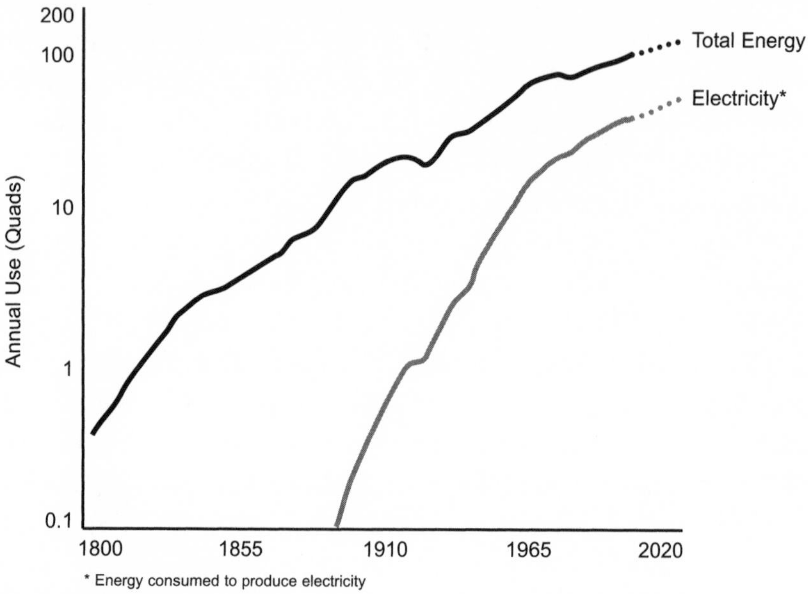

Source: EIA, Annual Energy Review 2003; U.S. Census Bureau, Historical Statistics of the United States Colonial Times to 1970.

U.S. consumption of Quads (raw energy) rose ten-fold in the twentieth century; electricity consumption rose thirty-fold.

steam into and out of the cylinder. It added open/shut digital timing to hot and cold, and in adding that, Watt changed the world.[3]

Invented over half a century before Watt's, Savery's engine of 1698 functioned by alternately injecting steam and then cold water into a cylinder—the engine used cold water to condense steam to form a partial vacuum that sucked water up a pipe. There was no piston; the valves on the cylinder were opened and closed by hand, sequentially. It worked—Savery's engine was being used to pump water out of British coal pits decades before Watt's—but all the metal had to heat up and cool down in each cycle, and this was terribly slow and inefficient. The piston-based Newcomen engine that followed wasn't much better.

Watt grasped that efficiency could be improved considerably if the main cylinder was kept hot all the time. That required moving the cold half of the cycle into a separate condenser. To do that, the flow of steam between the cylinder and the condenser had to be choreographed in synchrony with the up-and-down motion of the piston. Watt invented his regulator to automate the valves that separated the hot side of the engine from the cold; the engine could now cycle much faster, and thus produce much more power.

To compare the power of his steam engines with the power of the horses they displaced, Watt defined a new unit of measure, the "horsepower."* This was the useful work done when a horse lifted a 550-pound load of water or coal out of a mine shaft at a rate of one foot per second. Watt chose a mine shaft for a simple reason—the first and most important use of what would become the universal engine of the coal-fired industrial revolution was to facilitate the mining of coal.

JUMBOS AND JETS

Now suppose that Watt had been able to imagine every subsequent improvement others would bring to his thermomechanical engine, and all the things its progeny would end up doing. Not just steam-powered trains, but diesel trucks, jumbo jets, Ski-Doos, leaf blowers, Space Shuttles—every last engine that burns fuel to push things or people about—through a factory, down a highway, across the ocean, or through the air. Today, all of those applications together consume about 30 percent of all the fuel we burn. Even with that much prescience, Watt would still have missed over half of the new demand for energy that his engine would spawn. The engine that began moving coal would in due course be fired up to move electrons, and the electrons would turn out to be heavier. Electron-propelling "fire-engines" now consume 40 percent of our raw fuel.

Steam power met the electron in 1882, at Thomas Edison's Pearl Street Station power plant, in New York City. Edison had designed and built six "Jumbo Engine-driver Dynamos," each one a 27-ton, 100-

---

*When scientists found themselves in need of a new name for a unit of electrical power more than a century later, they too looked backward and settled on the "Watt."

kilowatt (kW) behemoth, four times bigger than any other electric generator previously built. The Jumbos were powered by high-speed coal-fired
steam engines. The entire useful output of all these tons of steel, and the
many thousands of tons of coal that they would burn, was channeled
down thin metal wire—15 miles of it, snaking through the city's bustling
financial district, leading to the eighty-five customers that had installed
Edison's new electric lamps.

Ironically, the electron had met steam power six years earlier. Building
on his experience with electroplating and railway telegraphs, the Belgian
Etienne Lenoir had designed an electrical ignition system for an engine
in which illuminating-gas burning inside a piston-cylinder replaced steam
produced outside it. Nikolaus Otto learned of Lenoir's engine and realized it could run much better on liquid fuel. He patented a two-stroke
engine design and won a gold medal for his invention at the 1867 World's
Fair in Paris. In May 1876, Otto built a four-stroke internal combustion
engine, the first practical alternative to the steam engine. Some thirty
thousand of these "Otto cycle" engines would be sold in the following
decade.

With an external furnace and condenser, Watt's steam engine was efficient by the standards of its day, but far too heavy for use in transportation. Early steam locomotives solved half the problem, by getting rid of
the condenser and simply venting steam to the atmosphere, but that
sharply lowered efficiency and required frequent stops to refill the boiler.
By putting the furnace, which converts fuel into heat, *inside* the piston-
cylinder that converts heat into motion, the "internal combustion" engine was much lighter and smaller than its "external" father. But Otto's
engine required twice as much synchronization as Watt's steam engine—
one timing system to move material in and out, as before; a second to get
the ignition timed just right. The ignition systems developed by Lenoir
and Otto supplied the additional logic.*

---

*Otto also developed the carburetor to control the mixing of liquid fuel and
air as it was drawn into the cylinder. Rudolf Diesel would later come up with a
third alternative, in which the piston compresses the fuel/air mixture to such a
high pressure and temperature that it ignites spontaneously, without a spark.

FIGURE 2.2   Aviation Engine Power Density

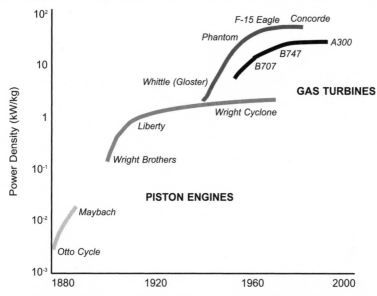

Source: Vaclav Smil, *General Energetics: Energy in the Biosphere and Civilization* (Wiley, 1991).

Piston engines supplied enough power per pound to get the Wright Brothers into the air. Supersonic jets required a hundred-fold increase in power density, which was supplied by gas turbines.

It took another half century to get rid of most of the remaining bulk of Watt's original design—the piston-cylinder itself. While Edison and Otto were electrifying lights and engines, the Swedish inventor Gustaf de Laval was working on a design for the first steam turbine, which he completed in 1882. A turbine is a remarkably simple and compact device —a windmill in a metal casing, with a stream of gas serving as the wind. De Laval invented a nozzle that accelerated steam jets to supersonic speeds, and patented the first functional steam turbine in 1883. But perfecting the device took many more years, because a turbine does not run efficiently until the blades rotate at close to the speed of sound. This requires advanced materials, very sophisticated machining, and high-speed bearings.

In a turbine, the Watt-engine logic of piston strokes and synchronized valves gives way to the much finer logic in the curvature of the nozzle

FIGURE 2.3   Engine Power Density and Shaft Speed

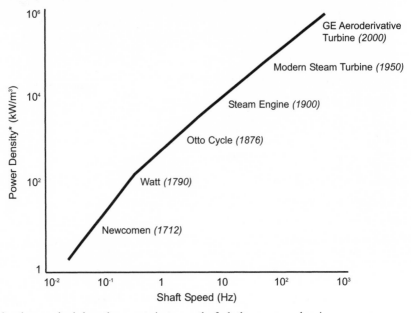

Engines have evolved along the same trajectory as the fuels they consume, burning more energy, faster, at higher temperature, to produce more power in less space.
*for engine only

and blades, which channel the continuous flow of very hot, high-speed gas and draw energy from it as they do. In effect, the cylinder, piston, and regulator are all incorporated into these two precisely shaped components, allowing the engine to produce far more power in far less space. The internal-combustion piston engine was enough of an advance over steam to get the Wright Brothers into the air in 1903. But the gas turbine was what made possible commercial air travel as we know it today. Turbines are now used wherever high power-density is essential—in jets, helicopters, war ships, and the Abrams M1 tank.

Engines have thus evolved along exactly the same trajectory as the fuels they burn. Generation by generation, they burn richer fuel, faster, hotter, in less space. One turn of the planet's thermal cycle, which fuels life in the biosphere, takes twenty-four hours to complete. Savery's steam engine could operate at a cycle or so every few minutes. Watt's regulator pushed engine speeds up to cycles per second. Otto's internal combustion engine, choreographed by an even more complex array of valves, made possible hundred-cycle-per-second engines. De Laval's turbine now runs at over a thousand cycles per second—hence the high-pitched whine of a jet engine. When de Laval died in 1913, his memorial was engraved with the inscription: "The Man of High Speeds."

From Watt to Otto to de Laval, the progress of combustion-engine technology can be mapped on a single curve of rising speed and power density, with power density increasing by a factor of ten or so in each generation. And each succeeding generation has allowed us to do new things that were simply not possible before. Watt's steam engine could produce far more raw horsepower, in far less space, than any manageable team of horses—making possible the mechanical loom and the "iron horse" locomotive, which together transformed all manufacturing and transportation. Demand for coal soared. Otto's internal combustion engine could produce far more raw horsepower, in far less space, than Watt's engine—making possible the "horseless carriage" and further, profound advances in industrial machinery. Demand for oil soared. De Laval's turbine marked another quantum leap in speed, power density, and rapid consumption of fuel. It gave us, among other things, the wings of Pegasus.

No government edict, no "energy policy," played any role at all in this history. These things happened because humanity hungers for power, speed, and range; because the technologies of power were invented; and because raw materials were then somehow found to fuel them.

ELECTRIFICATION

The generators that power Pearl Street's lights today are turned by 100 to 250 megawatt (MW) steam turbines, turbines a thousand times more powerful than Edison's Jumbos. Every large coal-fired plant is built alongside railroad tracks, so that trains can supply it with over 2.5 million tons of coal a year. Gas-fired plants, fed by massive pipelines, don't need trains to deliver their fuel, but they use jumbo-jet engines to burn it.

How coal might—in principle—be converted into electricity had been worked out by Michael Faraday in 1831.* Move a wire in a magnetic field, and you induce a current—that's what the Edison's Jumbo generators did in 1882. Or push a current through the wire, and you can move either the wire itself or the magnet—that's what electric motors were doing not long after.

But electric generators and motors require precise timing and logic. Magnets and coils of wires must be arranged so that—in one typical configuration—the current and voltage flip in precise synchrony with the flipping of the alternating current. It was a Croatian immigrant to the United States, Nikola Tesla, who developed the first practical electrical counterpart to Watt's regulator, and variations on this invention would become the linchpin of all generators and motors. George Westinghouse bought the patent in 1885 and used Tesla's system to light the World Columbian Exposition in Chicago in 1893. Today, about one-quarter of the electricity we produce is used by devices that collapse Tesla, Edison, and Watt into a single box and run them in reverse, with electricity being used to generate motion to move a hot gas to pump heat out of places where we don't want it, to cool things down.

Electricity is the most compact and the fastest form of power in widespread use—it pushes power density and speed far beyond the gas turbine that generates it. About 40 percent of all the energy channeled into hundred-car trains of coal, enormous gas pipelines, and building-sized gas and steam turbines in power plants ends up coursing at the speed of light through metal wires a few inches thick. The electrical power train is overwhelmingly superior to the mechanical—five orders of magnitude better on every key metric. Small wonder, then, that electric power has been steadily displacing all other forms, at the front end of our energy economy.[4] Today, about 60 percent of our GDP comes from industries and services that use electricity as their core "fuel"; in 1950, the figure

---

*While still chancellor of the exchequer, and before he became Britain's prime minister, William Gladstone visited Faraday's laboratory. "But, after all, what good is it?" he asked, when Faraday explained how he could transform motion into current. Faraday replied: "Why, sir, one day you will tax it."

was only 20 percent. The consumer pays a stiff premium for the ingenuity embedded in the highly complex hardware required to generate and distribute such highly ordered power.

THE POWER OF LOGIC

Many end-of-growth pundits concluded that this march from technology to technology, from steam to motive power to electricity, was set to end around 1980. Light bulbs had propelled a first great wave of demand for electricity, motors a second, and air conditioners a third. What then was left? Nothing much, it seemed. As we noted in the preface, one widely quoted pundit was quite sure that the future of electric demand was "as flat as the Kansas horizon."

He was wrong. Like most everyone else, the flat-power prognosticators utterly failed to foresee (among other things) the rise of the microprocessor and the Internet. We dwell on this for the next few pages only to illustrate how painfully easy it is, with energy, *not* to anticipate the new technology that will propel the next great wave of new demand.

Charles Babbage did anticipate it, but way too early. Barely fifty years into the life of the steam engine, it dawned on the great Cambridge mathematician that some form of energy other than his morning breakfast ought to be enlisted to help him do arithmetic. In the 1821 passage quoted at the beginning of this chapter, Babbage was referring to his friend and fellow astronomer John Herschel, and to tables of stars that had been commissioned the year before by the Astronomical Society of London. The hand calculations (performed by Babbage's colleagues, whom he called "computers") required enormous effort, and errors were infuriatingly common. Babbage was convinced that there had to be a better way, and he went on to design and build several huge mechanical calculators of remarkable ingenuity. They were all, however, operated by hand cranks—Babbage himself never got to the point of propelling his computers with steam.[5]

John Mauchly and John Presper Eckert did, in 1946. Their Electrical Numerical Integrator And Calculator—the ENIAC—was built to

FIGURE 2.4   Energy and Logic

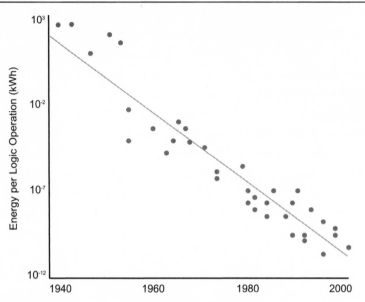

Source: R. Isaac, "Influence of Technology Directions on System Architecture," IBM Research Division, Sep. 10, 2001.

In 1946, the vacuum tubes in the ENIAC computer required 10 Watts or more to execute a single logic operation. Today's semiconductor gates are over ten million-fold more efficient.

compute tables for the trajectories of artillery shells rather than stars. The machine contained 17,468 vacuum tubes, along with 70,000 resistors, 10,000 capacitors, 1,500 relays, 6,000 manual switches, and 5 million soldered joints. It weighed 30 tons and consumed 174,000 watts (W) of electrical power—enough to brown out parts of the city of Philadelphia whenever it was fired up.

Today, steam-propelled electrons light all our countless microprocessors, memory chips, and video screens. A Pentium requires a power cache built right on to the semiconductor itself, to assure the steady flow of electrons to the gates. The individual gate is turned on or off by the addition or removal of a few hundred thousand electrons; that number is projected to fall to only a few thousand electrons by 2010. Could it drop to one? Or to none? We return to this very deep question toward the end of this book; suffice it to say that the answer is no, not to "none"—unless we are prepared to wait forever for the computer to spit out its answer.

For now, however, technology does keep pushing down the amount of power required to perform a single logic operation. The power consumed by a single gate in a microprocessor or memory chip depends on the size of the gate, and the gates keep getting smaller. Thus, the electrical energy required to process a single logic instruction is cut in half about every fourteen months. The ENIAC required about 10 watts per tube (i.e., per "gate"); a Pentium requires about one-one-millionth of a watt per transistor, which does the same job twenty thousand times faster. Similar rates of improvement in power-to-logic conversion efficiency have been realized with the storage of bits on a disk drive and their transmission through fiber-optic glass. The power requirement per bit falls; the "bit efficiency" rises.

But the smaller the gates, the faster we run them—that's one big reason for making them smaller in the first place. Gigahertz (GHz) microprocessor speeds—a billion clock cycles per second—are now routine in desktop computers. The power amplifiers in wireless communications systems are now pushing toward the tens of gigahertz. Millimeter wave radars are heading for the hundreds of gigahertz. Power electronics in optical telecom systems now run at terahertz (thousand-gigahertz (THz)) speeds. Comparable speeds (tens of megahertz (MHz) and up) are encountered in the electronics that drive optical displays, radar systems, and the best audio amplifiers.

And the number of gates per chip rises as fast as the gates themselves shrink, and faster still, as chip areas grow larger. The total power required by each chip thus rises even faster. An Intel engineer presents a remarkable forward-looking projection of where the power densities on the surface of a chip would be headed if current trends were to continue. Amusing though they were surely intended to be, the captions on his graphic sum up an altogether serious story: "current chip," "nuclear reactor," "rocket nozzle," and "sun's surface." Cooling microprocessors to keep them from melting down is already a major challenge.

Logic gates have power appetites so minuscule one can almost count the individual electrons that flow back and forth to flip each individual

FIGURE 2.5   Microprocessor Peak Power

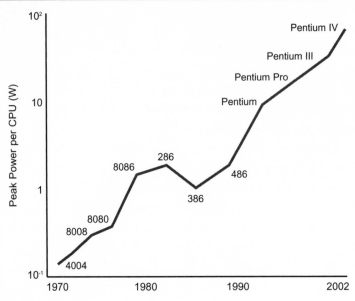

Source: Intel.

Total power per microprocessor rises because the number of gates per chip, and the speed at which they switch on and off, rise faster than the power-per-operation falls.

gate or store each individual bit. But the operation of these triumphs of miniaturization now depends on mirror-image arrays of power chips and power-caching capacitors, mounted all around the microprocessor, to assure the steady supplies of electrons that define data and conduct the logic calculations. The logic gates run at microamps and microwatts. A centimeter or so away, the power chips and capacitors have to serve up currents a hundred million times larger.

What do all the digital loads add up to, across billions of chips, phone lines, switches, routers, servers, data warehouses, and associated power quality hardware? In 1958, Hans Thirring quantified, almost certainly for the first time, how much energy might be required to power the hardware of telecommunications and medicine—his objective was to debunk "widespread misconceptions about the influence of radio waves on the climate."[6] Thirty years later, the National Academy of Sciences made the first serious attempt to quantify how much electricity was required to fuel the burgeoning infrastructure of information technology.[7] In an article published in *Forbes* in 1999, the authors of this book estimated that the manufacture and use of computers and networking hardware, along with their power backup and cooling systems, consumed some 8 percent of U.S. electric power;* the number rose to about 13 percent when we added in the rest of the information-technology infrastructure deployed at the foothills of the Internet.[8]

Our estimates triggered a furious response from a small but vocal academic circle, which maintains that the Web actually reduces energy consumption.† Our numbers were too high by a factor of four, or perhaps ten, these critics declared. Doonesbury caught up with the debate on October 23, 2000, by way of presidential candidate George W. Bush, who had cited one of our estimates in a speech.‡

A reasonably even-handed assessment of the fracas eventually appeared in an article by Brian Hayes in *American Scientist*.[9] Hayes reproached us

---

*In 1995 and 1997, according to estimates from the Department of Energy's Energy Information Administration (DOE/EIA), residential and commercial PCs alone accounted for between 4 and 6 percent of national electric use. DOE/EIA, *A Look at Commercial Buildings in 1995: Characteristics, Energy Consumption, and Energy Expenditures*, October 1998, and DOE/EIA, *A Look at Residential Energy Consumption in 1997*, November 1999. Our totals included the electricity required to manufacture digital equipment, to power wired and wireless networks, and to cool the digital hardware.

†One of the more vocal denunciations came from a group that had published a memorably wrongheaded conclusion about computers and power just a few years earlier. "The US commercial sector market is becoming saturated (especially for PC CPUs and monitors)." That assessment had been published just as the Internet began its explosive growth. "Efficiency Improvements in U.S. Office Equipment," Lawrence Berkeley Laboratories, December 1995.

‡W's disembodied hat floated above Trudeau's riposte: "'We're not talking fuzzy-wuzzy math here, folks . . . Well, not for me personably! . . . Even though the Internet uses 8% of our energy! Those don't count as lies, do they?' 'No, no. That's Gore. You're the dumb one.'"

FIGURE 2.6   Computing and the Internet: Aggregate U.S. Electricity Consumption*

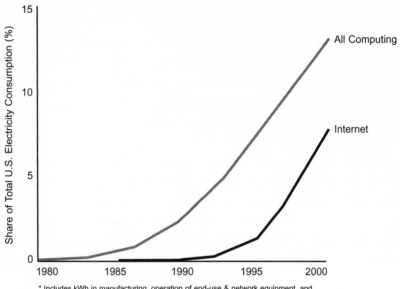

* Includes kWh in manufacturing, operation of end-use & network equipment, and infrastructure.

Source: Peter Huber and Mark Mills, "Silicon & Electrons," Dec. 2002, www.digitalpowergroup.com.

Billions of chips, hundreds of millions of phone lines, tens of millions of switches, routers, and servers, thousands of data warehouses, and power quality hardware add up to significant demand for electricity. The consumption totals shown include power consumed in manufacturing digital equipment, the operation of both terminals and networks, and cooling loads.

for insufficiently documenting our estimates—but ended his piece with a summary of data he had gathered on his own. Much to his surprise, he found that the computer equipment in his own home accounted for 9 percent of his electricity consumption. But "it would be foolish to extrapolate from my home office to the entire national economy," Hayes hastened to add, because as a journalist he "surely spend[s] more time at the keyboard than most people do."

Perhaps he does. But Hayes also neglected to consider the power used to manufacture all the equipment he was using; the additional air-conditioning loads in his home created by all his hot electronics; the power consumed by the telephone line he uses to access the Internet, or the considerably larger amount of power used by a DSL line or cable modem, if he's got one; the power the telephone company central office uses to serve his line; and the power used by AOL to serve up data to him when he wants it. And if Hayes himself spends more time at his home office keyboard than most, he probably spends less time at a downtown-office keyboard, or on an automated factory assembly line—where digital machines are used far more intensely than they are in even the most richly wired homes.

Hayes, like many others, concludes that the challenge is to build more efficient microprocessors and run them smarter so that they use less power. But the ENIAC-to-PlayStation trajectory suggests that improving computing efficiency is not likely to reduce demand for power overall. The trend-setting 70-watt PlayStation II incorporates ten thousand times the computing power of the ENIAC, and requires two thousand times less power. But one PlayStation per teenager adds up to much more demand for power than one ENIAC per planet.

Whatever the right number may be today, how much more power might all our logic machines require in the future? This much is certain: in neurons, microprocessors, cell phones, and fiber-optic lines, all flows of information are flows of well-ordered power. Information may hibernate in chemicals (DNA), or tiny magnets (on metallic disks), or pits in plastic (in optical disks). But when it's awake, information is power—packets of electrons or their quantum cousins, photons. Bits are defined units of energy that get ordered, sifted, herded, and propelled through living tissue, silicon, the airwaves, and tunnels of copper, coaxial cable, and glass. Ordered packets of information cannot form, replicate, persist, or be conveyed until they are rendered incarnate as ordered power.[10]

Equally certain is that it takes more power to move bits faster. A semaphore uses daylight, Morse's telegraph required a battery, and it takes a precisely tuned laser to send terabits of data down a strand of glass. Gigahertz-speed logic calculations on the surface of a Pentium require fantastically stable DC power of extremely high density—the tiniest interruption or imperfection in the power flowing to the chip wipes out everything and means blue-screen death for the computation under way. Truly chaotic power—"black body radiation"—can neither process nor convey any information at all.

In our times, the power of steam has thus been transmuted into an altogether new form, packets of power so tiny, so compact, so precisely controlled that we no longer call them "power"—now we call them "bits." But they are, in fact, packets of energy, no more or less, and we consume them voraciously. The "bit efficiencies" of our microprocessors and communication channels rise astonishingly fast, but the total number of bits processed and conveyed rises faster still. Babbage designed a bit-propelling tricycle, Intel's first chip was a moped, our desktop PCs are now SUVs, and we will all be driving logic tractor trailers soon, and then Space Shuttles. The bit engines supply data, entertainment, connection, and control, and in due course they will supply insight, understanding, knowledge, intelligence—eventually, perhaps, even wisdom. It seems unlikely we will ever stop wanting more of *that*.

THE LOGIC OF POWER

James Watt's regulator controlled the flow of steam; the revolutionary new regulator of our era is the semiconductor gate, which controls the flow of electrons and photons. William Shockley, Jack Kilby, and Robert Noyce built the first such device in 1949, and called it a transistor. A decade later Kilby and Noyce thought to build resistors and capacitors alongside, to create an "integrated circuit" on a single semiconductor crystal. The size and the amount of power switched by these logic transistors has since been cut in half about every two years, with the amount of power that they switch falling apace.

But could the same device be pushed *up* the power curve instead, to switch—and thus impose order upon—the energy that assembles a car in a factory, or propels it down a highway, or powers the driver's house? In 1982, Hans Becke and Carl Wheatley (both at RCA) were granted a patent for a "Power MOSFET with an Anode Region."[*] The device,

---

[*] "The present invention . . . relates to vertical, grooved MOSFETs (VMOS) used in power applications. . . . The improved forward conductance . . . makes such a device particularly suitable for high voltage switching applications."

later named the Insulated Gate Bipolar Transistor (IGBT), combined hundreds of transistors on a single crystal. It was able to control large flows of electrical power as quickly and efficiently and cheaply as logic semiconductors could control bits.

The Becke–Wheatley "power chip" and its progeny exploit the same quantum-physical phenomena that are harnessed by integrated circuits and laser-driven fiber-optics. But now the semiconductor junctions, the microwave beams, and the lasers are handling hundreds of watts, or thousands or millions, to control the propulsion of heavy things like high-speed trains, trucks, cars, and industrial machines. They are moving and processing material in the blue-collar tiers of industrial enterprises, just as their older but much weaker siblings move and process information in the white-collar tiers. They project beams of photons that heat and cook and cut through steel, rather than just glide in frictionless comfort through ultra-pure glass.

The new technologies begin with materials that are plucked from favored columns of the Periodic Table: silicon, gallium, germanium, carbon, arsenic, nitrogen, indium, phosphorous, aluminum, boron, tin. They are then united, with fantastic purity and precision, in an alphabet soup of compounds and junctions, a babble of material combinations never imagined or used in any branch of engineering until the advent of semiconductors and the vast infrastructure of semiconductor tools and machines that purify, assemble, and machine them.

In their fundamental operation, semiconductor devices exploit bizarre quantum effects that are manifest only in atomic-scale structures. Because their functional parts are so small, these devices are extremely fast and efficient—orders of magnitude better than all the switches, filaments, antennas, and sensors they are now rapidly displacing. These devices permit such accurate control of power that we can now, for the first time

---

Hans W. Becke and Carl F. Wheatley Jr., U.S. Patent 4,364,073 (December 14, 1982). A similar device was described by B. J. Baliga of General Electric the same year. See B. J. Baliga et al., "The Insulated Gate Rectifier (IGR): A New Power-Switching Device," in *IEDM Technical Digest*, 1982, pp. 264–267 (International Electron Devices Meeting in San Francisco).

FIGURE 2.7   Transistors for Logic and Power

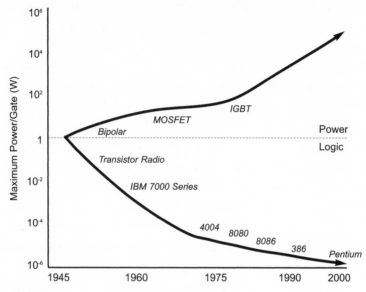

Source: Intel; IBM.

The power "regulator" of the digital age is the semiconductor gate that controls a flow of electrons or photons. William Shockley's team built the first transistor in 1949; the power switched by logic transistors has since been cut in half about every two years. Pushing semiconductors up the power curve took longer. It wasn't until 1980 (patent issued in 1982) that Frank Wheatley and Hans Becke of RCA invented the Insulated Gate Bipolar Tansistor (IGBT), capable of switching kilowatts in industrial robots, hybrid cars, aircraft, trains, tanks, ships, and machine tools.

ever, speak of *digital power*—power under the control of systems so fast and precise that the power becomes as tractable and ordered as digital logic.

From these new classes of semiconductor devices, manufacturers are now building ultra-compact and efficient power supplies, capable of shaping voltage and current in any manner desired, to power a microwave oven, microprocessor, laser, and all of the rapidly multiplying layers of electrical grids, motors, sensors, and actuators in satellites, jets, tanks, telephone company central offices, and wireless base stations. With complete control of voltage and frequency, it is now possible to build ultra-compact electric motors—motors as powerful as a car engine but no bigger than a coffee can.

New families of semiconductor-based radios, lights, and lasers are emerging in tandem. They are the technologies that are transforming industrial heating; they are likewise transforming all forms of communication, lighting, and electronic vision. Light-emitting diodes (LEDs) are already being built by the billions out of aluminum-gallium-arsenide, gallium-nitride, and other semiconductors. Silicon and then gallium-arsenide power-chip amplifiers made possible the cell phone revolution. Indium-phosphide amplifiers will soon make car radar systems as standard as headlights. In our cars and elsewhere, we are fast moving toward perfect vision—systems that can penetrate fog, foliage, flesh, plaster walls, and other forms of clutter, and detect not just shape and color but internal structure and even molecular composition, because they shine and see not just visible light, but microwaves, millimeter and terahertz waves, infrared, ultraviolet, and X-rays. A decade ago, no one would have imagined such capabilities coming to the mass market. They will soon be ubiquitous.

These technologies of digital power now make possible an extraordinary new variety of compact, affordable, product-assembling, platform-moving, people-moving, and power-projecting systems that seem all but magical. In the coming decade, they will infiltrate, capture, and transform the capital infrastructure of our entire energy economy—the trillions of dollars of hardware that convert heat into motion, motion into electricity, and ordinary electricity into highly ordered electron and photon power. They will transform the factory floor, the hospital operating room, and every aspect of civil defense and warfare. And they will transform the automotive sector, whose massive demand for components themselves, and component-manufacturing technologies, will push down prices and thus propel these same technologies into other sectors.

They are, in short, as revolutionary as Watt's steam regulator was in 1763, as Otto's spark-ignited petroleum in 1876, as Edison's electrically heated filament in 1879, as de Laval's hot-gas turbine in 1882. And they too will redefine, yet again, how much energy we want and how much we can get. We will want more—much more. And we will get it, easily. Unless, somehow, our optimism, drive, courage, and will give way to lethargy and fear.

3

THE VIRTUE OF WASTE

Nature, in providing us with combustibles on all sides, has given us the power to produce, at all times and in all places, heat and the impelling power which is the result of it. To develop this power, to appropriate it to our uses, is the object of heat engines. The study of these engines is of the greatest interest, their importance is enormous, their use is continually increasing, and they seem destined to produce a great revolution in the civilized world.

—SADI CARNOT, REFLECTIONS ON THE MOTIVE POWER OF HEAT AND ON MACHINES FITTED TO DEVELOP THAT POWER (1824)[1]

TO PUT IT as bluntly as it can be put, the "waste" of energy is a virtue, not a vice. It is only by throwing most of the energy away that we can put what's left to productive use. The cold side of the engine—where we discard most of the energy—is as essential as the hot, where we suck it in. More essential, in fact. It is by throwing energy overboard that we maintain and increase the order of our existence.

Consider, for example, the magnificent laser. There is without doubt something fascinatingly wonderful about the beam it produces—so cold, yet so intense. On the most simple-minded metric, sunlight and laser light are exactly the same thing—streams of photons. In terms of the order, however, they are almost as different as night and day. One supplies cow-pasture-quality energy; the other supplies energy as dense as may be found in the core of a nuclear reactor. With junk photons you can grow grass; with highly ordered photons you can grow semiconductor chips in a fab, where the deep ultraviolet laser is the most essential part of the lighting. Junk photons can kick electrons a few volts up the energy ladder on the photovoltaic semiconductor in a solar cell; highly ordered photons can punch holes through steel.

There is something disgraceful about a laser too—on any simple-minded accounting of energy consumption it is the quintessence of inefficiency. A laser burns light to generate light. The wave-like oscillations of the laser beam are all in step because light bouncing back and forth between mirrors in a cavity is used to stimulate the emission of more light. In gas lasers, the input light is pumped into a large, evacuated gas chamber; in solid-state lasers, it is pumped into a much smaller space constructed within a crystal. Either way, the device burns light to generate light, discarding most of the inbound energy in the process.

The laser's miserable plug-to-beam efficiency is only the last chapter in a long saga of dissipation. Directly behind the laser stands a complex array of power electronics—electrical engineers call it a "power supply," though it consumes power in the process of supplying it. This tier of waste is required to convert relatively noisy, unreliable electricity from a wall plug into a form that is pure enough to run the laser.

The electricity at the plug arrives there from the enormous generator in some utility's central power plant. What spins the generator's shaft is a steam turbine. The steam comes from a boiler, which is heated by furnace, which most probably burns coal. In the very best power plants, half of the raw heat available in the coal is consumed inside the plant itself, in converting the other half of the heat into electricity. Less efficient power plants—smaller ones used as stand-by generators, for example—consume two-thirds of their heat to refine the other one-third into electricity. The whole business, in short, reeks of a Ponzi scheme, with each successive tier of the pyramid feeding voraciously off the one beneath—and with new tiers constantly being added at the top. Small wonder that so much of our energy economy is often characterized as wasteful. Casual observers are easily convinced that there must be a better way.

FIGURE 3.1   Pyramid of Energy: Bits and Photons

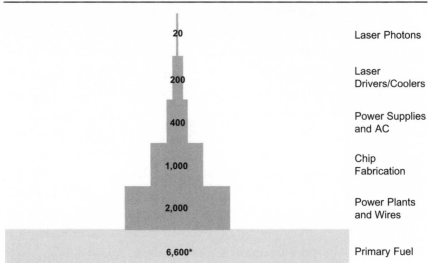

20	Laser Photons
200	Laser Drivers/Coolers
400	Power Supplies and AC
1,000	Chip Fabrication
2,000	Power Plants and Wires
6,600*	Primary Fuel

* 6,600 kWh thermal energy is roughly 4 barrels of oil.

A laser's intensely ordered flow of photons is far more useful than the sunlight that grows grass. But the laser beam depends on complex arrays of generators and power electronics behind it, which dissipate most of the energy in the raw fuel in the process of converting a tiny fraction of it into perfectly ordered photons.

The energy Ponzi scheme is invariably framed—and lamented—as a symptom of grotesque *waste*. In the standard graphical presentation, the noble pyramid is portrayed, instead, as a squid-like creature, expelling waste through every tentacle. The first such graphic was apparently drawn in 1949;[2] it brings to mind another perhaps more familiar graphic depicting the progressive destruction of Napoleon's army on its march to Moscow and back.[3] Updated versions of the energy squid are now routinely wheeled out to demonstrate how most of the energy we use goes to "waste"[4] or (more colorfully) disappears down a "rat hole."[5]

But something far bigger than a wasteful rat hole is at work when you are looking at 95 percent or more of total demand. *That* much demand can't all be blamed on bad engineering. If the main use of energy is to condition energy itself, then "energy" isn't the right metric at all, and the "energy economy" must in fact center on something quite different.

FIGURE 3.2 Origin and Utilization of the World's Energy in 1937

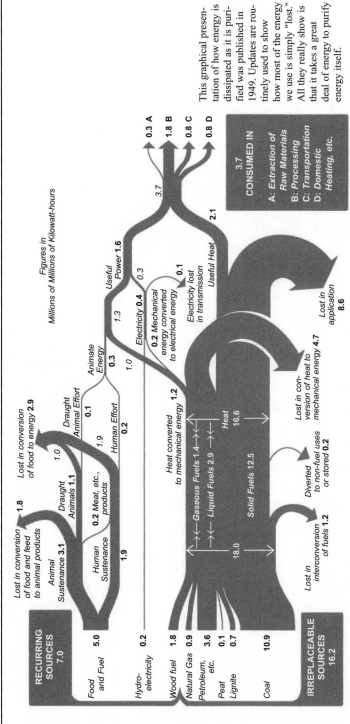

This graphical presentation of how energy is dissipated as it is purified was published in 1949. Updates are routinely used to show how most of the energy we use is simply "lost." All they really show is that it takes a great deal of energy to purify energy itself.

Source: N. B. Guyol, Energy Resources of the World, Department of State Publication 3428 (U.S. Government Printing Office, June 1949), reproduced in Hans Thirring, Energy For Man: From Windmills to Nuclear Power (Harper & Row, 1958), p. 43.

Engines and generators are obviously doing something for us that isn't captured by any of the conventional metrics of energy and power.

In fact, the huge pyramid (or squid) of consumption, with its withering "losses" at every turn, doesn't reflect bad engineering; it reflects, in all its real-world complexity, one of the most fundamental laws of physics, the second law of thermodynamics. The purification of energy depends on very complex structures that turn energy inward on itself, so that two units of low-grade energy funneled into the machine emerge as one unit of high-grade energy and one unit of useless heat. Energy doesn't just lounge about waiting for the chance to propel moms and kids to soccer fields—getting things to that point is an uphill battle. The remarkable thing isn't that our power-conversion technologies are inefficient but that they work at all.

What they are doing is purifying power. Most of our demand for energy derives from energy's capacity to refine energy, and from power's capacity to purify power. *Our main use of energy—by far the most important in the "energy" racket—is to purify energy itself*. It is only by grappling with that strange fact that we come to understand why we use so much energy, and why we will always demand still more.

PYRAMIDS OF VIRTUOUS WASTE

Thomas Edison's first light bulb wasn't at all efficient. One 1905 observer complained that "the incandescing lamp is an extremely poor vehicle for converting electric energy into light energy, since only about 4 percent of the energy supplied to the lamp is converted into light energy, the remaining 96 percent being converted into heat energy."[6] And the power plant that Edison built to light his bulb didn't convert even 10 percent of its heat into electricity. But the end-to-end losses of over 99 percent seemed worthwhile to produce such a wonderfully clean, compact, cool, and safe source of light. Efficiency was beside the point. As Jill Jonnes recounts in *Empires of Light*,[7] gas and oil lamps didn't stand a chance against such a superior alternative.

Edison's generators were powered by steam engines—direct descendants of James Watt's engine developed a century earlier. Watt's early designs achieved no better than 5 percent heat-to-horsepower conversion efficiency—95 percent of the available thermal energy was dissipated in converting the other 5 percent into the torque and spin of a shaft. Even that was a lot better than was achieved by the Savery and Newcomen steam engines that preceded Watt's. Better than the Space Shuttle, too, if you do the bookkeeping honestly. Gargantuan quantities of fuel are used to get the Shuttle's payload into orbit; almost all of it is consumed in the very earliest stages of the flight, to lift and propel the fuel still on board, in yet another Ponzi scheme of energy consuming energy.

Yet here, too, the overhead has been considered a bargain. Watt's customers recognized from the get-go that the new "horsepower" supplied by Watt's engine was much better than the old power supplied by a horse and its feedbag—steam could do things that horseflesh could not begin to match. NASA's rocket designers know where all their fuel is going, but are determined to launch the Shuttle anyway, and nobody has yet devised a better way to turn liquid hydrogen into a couple of tons of useful payload orbiting the earth at 16,500 mph.

Finally, beneath Watt's steam engine there's the coal itself, along with yet another slice of overhead. It takes energy to extract the fossil fuel from where it lies haphazardly buried in the earth. It takes even more energy to extract oil, and still more to refine it. Up to 5 percent of the energy in the crude is used up separating low-grade tar from high-grade gasoline.

These pyramids of demand are universal, and they all look much the same: massive amounts of low-grade energy are consumed to deliver relatively tiny amounts of high-grade power. It takes a great deal of raw fuel to get a kilowatt-hour (kWh) of ultra-reliable electricity into a microprocessor CPU. A mom and kids in an SUV require very-high-quality power, too, and they get it—from a big engine that gets them places fast and lets them accelerate out of harm's way, along with a heavy frame to support the big engine and to shield them from what happens when high-speed energy turns abruptly into chaos, as it does when the SUV hits a tree. Only about 2 percent of the energy that starts out in an oil pool

FIGURE 3.3   Pyramid of Energy: Bits and Electrons

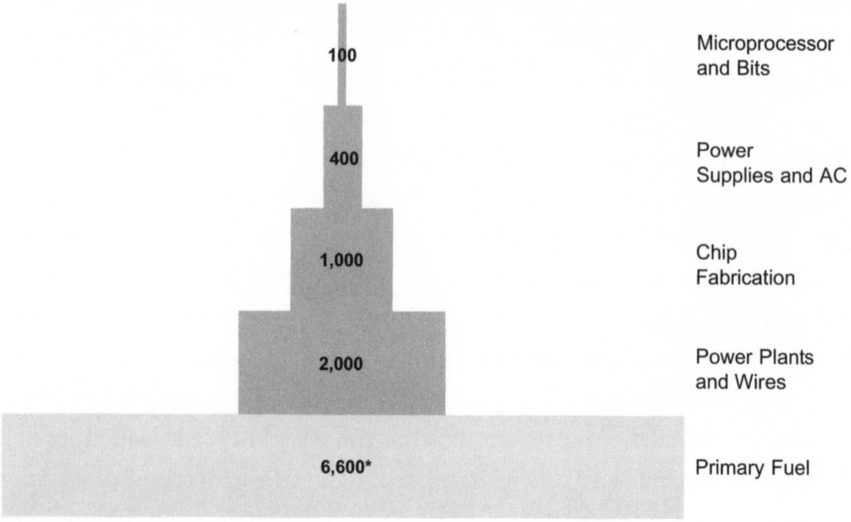

100	Microprocessor and Bits
400	Power Supplies and AC
1,000	Chip Fabrication
2,000	Power Plants and Wires
6,600*	Primary Fuel

* 6,600 kWh thermal energy is roughly four barrels of oil.

A great deal of raw fuel is needed to get a trickle of ultra-reliable electricity into a microprocessor CPU. A rapidly growing share of our electricity is now used to transform ordinary grid electricity into computer-grade power.

2 miles under the Gulf of Mexico ends up propelling 200 pounds of mom-and-the-kids—the ultimate payload—2 miles to the soccer field.

Life itself has evolved as an energetic pyramid. As Paul Colivaux discusses in his 1978 book, *Why Big Fierce Animals Are Rare*, plants are abundant, but they are neither efficient nor systematic in how they capture energy from a very diffuse source—sunlight. Herbivores capture perhaps 10 percent of the energy in the comparatively energy-rich plants on which they graze and browse. Carnivores extract a somewhat larger fraction from the herbivores that they manage to capture. Each successive level contains species that are roughly ten times larger than those in the level below, but also ten times less numerous. This "grand pattern of life" was first described by Charles Elton of Oxford in 1927: "the animals at the base of a food-chain are relatively abundant, while those at the end are relatively few in numbers, and there is a progressive decrease in between the two extremes."

FIGURE 3.4   Pyramid of Energy: SUV

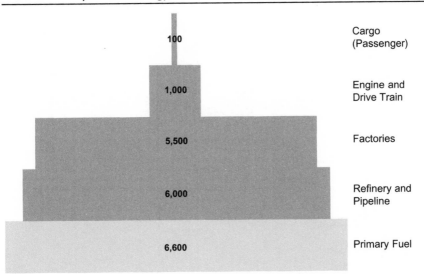

Energy consumes itself at every stage of its own production and conversion. Only about 2 percent of the energy that starts out in an oil pool two miles under the Gulf of Mexico ends up propelling two hundred pounds of mom-and-the-kids two miles to the soccer field.

Energy thus consumes itself at every stage of its own production and conversion, from the grassland on the Serengeti to the gazelle to the black-maned lion of Ngorongoro crater, from strip mine and derrick to the power plant and car engine, and from the direct current (DC) power supply to the central processing unit (CPU). Not just a bit of energy, here and there, but most of it. Over two-thirds of all the fuel we consume gets run through thermal engines—and well over half of it never emerges as shaft power at the other end. Just over half of all the shaft power we produce is used to generate electricity—but another 10 percent of that power doesn't make it out the far end of the generator. A rapidly growing share of our electricity is now used to transform ordinary grid electricity into computer-grade power—with another 10 to 20 percent overhead in this stage of conversion. Some small but growing fraction of high-grade electric power is used to produce laser light—and another 60 to 90 percent, or more, of the electric power dispatched to the laser never makes it into the blinding beam of light. These losses compound from end to end:

FIGURE 3.5   Pyramid of Energy: Life and Food

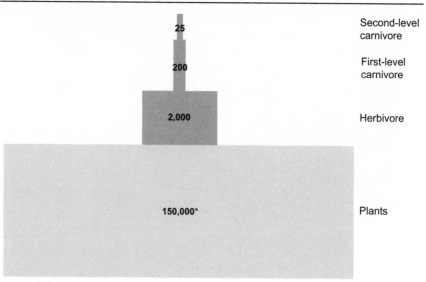

25	Second-level carnivore
200	First-level carnivore
2,000	Herbivore
150,000*	Plants

* Btu/m²/year (150,000 Btu roughly the energy in one gallon of gasoline)

Source: George B. Johnson and Peter H. Raven, *Biology*, McGraw-Hill (6th ed.), Figure 28.13c.

Energy is both purified and dissipated as it moves up the food chain in the biosphere. High-level carnivores depend for their survival on a huge expanse of plant life three steps below.

overall, only 1 to 5 percent (at best) of the thermal energy locked up in the fossil fuel or the enriched uranium ever emerges at the other end of the pipeline, as a laser beam, or a stream of cool air from an air conditioner, or as 200 pounds of 40 mph mom-and-kids; all the rest goes into purifying, conditioning, and tailoring the power.

ORDER

Energy is easy to grasp only insofar as energy is irrelevant; what's important, as Feynman observed, is *order*. Order is the "very subtle aspect of energy" that defines the difference between concentrated and dilute forms, and "it is very, very difficult to get right." One can say that it takes energy to move a car; one can say, with equal accuracy, that it takes energy to make the motion *stop*. One can likewise say that it takes energy to cool a room—yet it is a surfeit of energy that makes us want to cool it in the first place. "Energy" itself thus explains nothing; as noted earlier, we can't even "consume" it—energy is always conserved. The first law of thermodynamics says so. What matters isn't the energy but the order.

Thermodynamics gives us a rigorous measure for order—not a mere concept or metaphor, but a measure as quantitative and precise as temperature: that measure is *entropy*. Countless philosophers, psychologists, sociologists, and professors of English literature think they know what the word means, and invoke it to explain the decay of language, social order, global environment, and much else besides. J. Willard Gibbs, the nineteenth-century physicist—who really *did* know what the term meant —described entropy as "mixed-up-ness": it is a measure of disorder. The word itself is derived from the Greek for "in-turning"—the idea is that order tends to spiral inward upon itself, and turn into chaos.

But what, if anything, does *that* tell us? Most people have an intuitive feel for what order and chaos are all about. Few, however, have any idea how order can be systematically measured, or why it matters, or how they might buy a pint of order as a chicken-soup antidote to the muddle in their lives. They're pretty sure, however, that they can't buy it at a filling station. Energy, by contrast, is something they think they can grasp. Even many analysts think nothing of totaling up all energy consumption on a single thermal calculator—which means converting high-grade kilowatt-hours back into thermal units, and thus equating electricity to coal.

Which means that most people get things exactly backward. Again— energy is easy to grasp only insofar as energy is irrelevant. It is energetic *order* that matters, not energy itself. And our main use of energy—the use that completely eclipses all others—is to refine, process, and purify energy itself.

As we will discuss throughout this book, we use well-ordered energy to distance ourselves from chaos. But at the very outset, most of the energy is used to distance us from the chaos in energy itself. We begin that process with the only slightly better-than-chaotic, high-temperature heat that we create in burning raw fuel. And from that humble starting point, we climb the pyramid of energetic order until we reach the exquisitely choreographed microprocessor or the densely packed power of the ice-cold laser. Raw energy is nothing; context and order are everything; and all of the important action happens in getting from raw fuel far away to ordered power in close.

THE SECOND LAW[8]

Scientists and engineers know all this not just vaguely or intuitively—they know how to measure entropy exactly, as exactly as temperature, the very definition of which is inseparable from the definition of entropy. Entropy is indeed measured in units of energy divided by temperature—though knowing just that won't much clarify anyone's thinking about energy, efficiency, engines, or anything else. Suffice it to say (by way of example) that laser light is superb because it has very low entropy, which is another way of saying that it is fiercely hot, even though it is also, to most ordinary appearances, ice-cold—it knifes through the air, for example, without heating up the room. Or suffice it to say, instead, that there is no *energy* crisis. An *entropy* crisis, yes—*that* crisis is permanent, and forever getting worse. But not an energy crisis.

This subtle but fundamental truth about the universe was discovered over two centuries ago, by a young man who hadn't set out to study the universe at all. What irked this young Parisian was that British engineers had managed to push their steam engines so far ahead of their French counterparts, and without even bothering to work out a grand theory of what was going on. This offended his Gallic pride, and challenged his Gallic genius.

So Sadi Carnot set out to determine just how much useful work could, in theory, be extracted from a kilogram of steam. In 1824 he published his slim book on steam engines. It ran less than 120 pages. Only 600 copies were printed. But it set out, for the first time, what we now call the second law of thermodynamics, one of the most fundamental laws of the universe. Musing over the rankings of various scientific theories, Einstein concluded that classical thermodynamics is "the only physical theory of universal content which . . . within the framework of its basic notions, will never be toppled."

In the abstract, the second law is not at all easy to understand. One of its several formulations—the simplest, but also one that is wholly uninformative—is that the state of any "closed" system inevitably decays from more ordered to less. A second, almost as simple, but even less useful, states that heat flows only downhill, from hotter to colder, never uphill—unless you add something far better than heat—"work" it's now called—to spin the compressor in a refrigerator, for example. But these are all qualitative statements; the second law is in fact rigorously quantitative, and the formulations quickly get complicated.

To get any sort of grip on it, it is far easier to start where Carnot himself started, with a real-world engine. The key thing that Carnot grasped was that the condenser—the *cold* side of Watt's steam engine—was as important as the *hot*: it is the temperature *difference* that lets you extract useful work, and the bigger the gap in temperature, the more useful work you can extract. Push the hot side up to infinite temperature, or the cold side down to absolute zero (or do both, for overkill), and you can extract an infinite amount of work from an infinitesimal quantity of steam. And out of an infinitesimally small engine, too, if you can maintain this infinite temperature difference across an infinitesimally thin interface.

One half of this story seems pretty intuitive—the hotter you make the *hot* side of an engine, the more useful work you can extract from it. This (roughly speaking) is why it's difficult to extract a lot of useful energy out of any engine (or solar heater, or any other such device) that's heated by sunlight—the hot side just doesn't get very hot. Burn coal or oil, and you can extract far more useful work out of far less hardware. Thus, mechanical engineers push the temperature of engines up and up—the practical limits are determined by the materials at hand. A steel engine can run only so hot before the metal melts. A ceramic engine can run a lot hotter, and thus a lot more efficiently, and pack more power into less space, so long as your seals and lubricants hold.

FIGURE 3.6   Power Density and Energy Waste

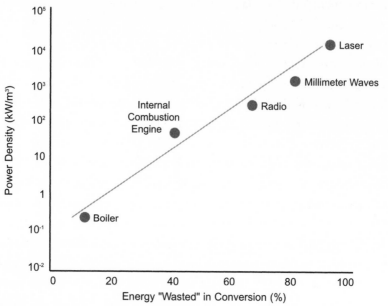

A system sheds "entropy"—chaos—only by shedding energy itself, in the form of waste heat. To produce more well-ordered energy, faster, in less space—to increase "power density," roughly speaking—one must throw away more energy, faster.

But the second half of Carnot's story, though of exactly equal importance, is far from intuitive. The *cold* side—the *condenser* of the steam engine—is exactly as important as the hot. The colder you make the cold side, the more useful work you can extract. A bath-temperature boiler generally can't run a powerful engine—unless you place a really, really cold condenser on the far side, in which case bath water will run the engine as powerfully as a white-hot flame of exploding gasoline. By the same token, you can't get any energy at all out of the white-hot flame unless you have a lower-temperature energy sink, drain, or dumping ground on the other side. What matters isn't how hot it is—that is, how much *energy* there is—in the boiler. What matters is the temperature *difference* between the hot side of the engine and the cold, and the carefully orchestrated *flow* of energy across this divide.

To put it yet another way—absolutely rigorous if wholly counter-intuitive—what permits order to increase is not the input of high-grade energy, but the dumping of low-grade energy, or chaos, to the surroundings. To increase the order of anything—including the order in energy itself—you *must* dump waste energy out the back door. A system sheds "entropy"—chaos—only by shedding energy itself, in the form of waste heat. To produce more well-ordered energy, faster, in less space—to increase "power density," roughly speaking—one must throw away more energy, faster.

This point is so fundamental, and explains so much about our consumption of energy, that it bears repeating. *To structure, organize, move, and increase order—of anything, anywhere—you have to add high-grade energy at one end, and then discard some fraction of it in the form of low-grade heat at the other.* There are no exceptions. This is how it works, always and everywhere.

Which brings us back to where we started. Waste is as virtuous as order, as virtuous as a tidy room, clean dishes, plaque-free teeth, a sterile operating theater, or ice in the refrigerator. You cannot get or maintain such things without dissipating high-grade energy, without dumping heat, without burning coal, oil, uranium, or wood, or finding some other equivalent means to capture quality energy and then dispose of it as junk. Life and growth being inescapably "dissipative" processes, waste is as virtuous as life itself.

4

SAVING THE PERILOUSLY
EFFICIENT GRID

We will make electric light so cheap that only the rich will be able to burn candles.

—THOMAS A. EDISON, ANNOUNCING HIS
ELECTRIC LIGHT BULB (1879)[1]

The subject which I now have the pleasure of bringing to your notice is a novel system of electric distribution and transmission of power by means of alternate currents, affording peculiar advantages . . . which I am confident will at once establish the superior adaptability of these currents to the transmission of power.

—NIKOLA TESLA, IN A SPEECH TO THE AMERICAN INSTITUTE
OF ELECTRICAL ENGINEERS (1888)[2]

BLAME SADI CARNOT for the great blackout of August 14, 2003, the one that darkened much of the northeastern United States and southern Canada. A blackout of that scale could occur only because the grid got built as it did. And it got built that way because Carnot's second law of thermodynamics decreed that it should be.

The second law establishes that the amount of useful work you can get out of a furnace depends on how much hotter things are in the furnace than they are on the cold side of the engine—the cold side being the

58

condenser, cooling tower, or whatever else is at hand to use as a dumping ground for the inevitable (and virtuous) waste. One way or another, you have to use the environment as the thermal dump, which generally means that the cold side of your engine has to be at the ambient temperature of the air or the nearby river. So all the gains in thermodynamic efficiency have to come from making the hot side hotter. Which means that you want to burn your fuel—your coal, uranium, gas, oil, or whatever else may be at hand—not just hot, but hotter, fiercely hot, as hot as you can possibly make it without melting down or blowing up all your expensive hardware.

Which means that to burn fuel more efficiently in a stationary power plant, you build a bigger furnace. Bigger systems are easier to keep hot because they have less surface per unit of volume, and because they can be surrounded by materials like concrete and steel that can both contain and survive the heat. There is, of course, much more than that to engineering efficient power plants. But first and foremost, the rule is simple: bigger can be hotter, and hotter is more efficient. So, decade by decade through the first century of electricity, power plants grew bigger, and in so doing grew more efficient.

The direct result, as noted in chapter 1, is that electricity has grown steadily cheaper in the 125 years since Thomas Edison first turned on the lights on Wall Street. Improved engine designs have further improved efficiency—turbines replaced piston engines, for example. And the turbines have grown colossally huge, which makes them not only more efficient but cheaper per unit of generating capacity. There are enormous economies of scale in building and maintaining these behemoths. Several thousand central power plants in the United States burn 30 percent more fuel than 200 million car and truck engines and run about three times as efficiently, producing almost four times as much useful power.

THE PERILOUS GRID

A one gigawatt (GW) plant—of which there are quite a few—can power the homes, workplaces, and factories of 400,000 people, but the power has to end up even closer to the end users than their cars. To get there, it moves either above the tarmac or underneath it. Measured by route miles and physical footprint, the North American grid is by far the largest network on the planet, aside from the roads and highways themselves. Generating stations dispatch electrical power through some 680,000 miles of high-voltage, long-haul transmission lines, which feed power into 100,000 substations. The substations dispatch power, in turn, through 2.5 million miles of local distribution wires to our toasters, computers, and industrial robots.

Blame thermodynamics, again, for making possible the relentless sprawl that created this continent-spanning ganglion of power lines. If we had to drive 150 miles to reach a gas station, we wouldn't bother—the round trip would consume a full tank of gas. But high-voltage electricity is such a dense, pure form of power that it can be dispatched over enormous distances with relatively modest losses, much as a beam of laser light can circle the globe through a sufficiently pure strand of glass. So, year by year, stretched out by the rising efficiency of swelling power plants, the wires have grown longer, and the average distance between where power is generated and where it is used has risen inexorably.

Though their electrical resistance does cause some losses, longer wires create an additional efficiency as well. Local demand for electricity varies a lot by time of day. When wires are long enough, the same power plant can accommodate peak demand at 4:00 P.M. in New York, and then a second time an hour later in Chicago, one time zone to the west. With lots of plants knitted together in a huge grid, all can operate at closer to full capacity for more hours of the day, which keeps things hotter still.

Finally, our own deep-seated aversion to chaos—our atavistic antipathy to entropy, one might say—has done much to stretch out the power lines even farther. Environmental and zoning regulations make it increasingly difficult to locate smokestacks and cooling towers near where we live. One of electricity's best features is that it *can* so readily separate the hot, dirty business of burning coal (say) from the cascade of sparkling ice from the door of the refrigerator. But the upshot of not-in-my-backyard

FIGURE 4.1   The Multitiered Grid

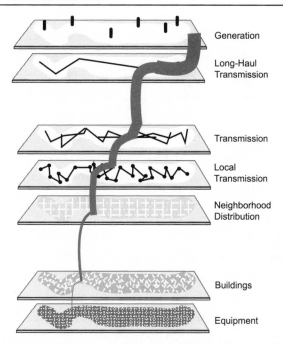

Measured by route miles and physical footprint, the multitiered North American grid is the second largest network on the planet, after the roads and highways. Generating stations dispatch electrical power through some 680,000 miles of high-voltage, long-haul transmission lines, which feed power into 100,000 substations. The substations dispatch power, in turn, through 2.5 million miles of local distribution wires to our toasters, computers, and industrial robots.

regulation is invariably a longer hike to some other less populated backyard over the horizon.

By every measure but one, these trends have been good. Over the long term, electricity has grown extraordinarily cheap because the sprawling grid is so very efficient. Its environmental costs—whatever they may be—are largely invisible: the Victorian home's filthy coal furnace has disappeared from our sight, along with its smoke. But with all this advantage has come one extraordinary, perhaps intolerable, peril. A structure that large, that sprawling, that exposed, is vulnerable to all sorts of assaults. And blackouts can be as far-flung as the vast network of wires that connects everything together.

ASSAULT ON THE WIRES

The forces of dispersion and decay do not readily submit to the grid's gigawatt-scale order. Weather has caused four massive outages in recent memory: Hurricane Andrew in Florida in 1992, Hurricane Fran in Virginia in 1996, and ice storms hitting the East Coast during the winters of 1998 and 2002. Spasms of human stupidity have worked their mischief too. In April 1992, construction workers installing support pillars in the Chicago River punctured the roof of a freight tunnel beneath the river bottom; the ensuing flood shut down utility power for weeks in the heart of Chicago.

When the grid does begin to fall apart, the collapse can propagate as far as the wires extend, and as fast as power moves, which is close to the speed of light. A single faulty relay at the Sir Adam Beck Station no. 2 in Ontario, Canada, caused a key transmission line to disconnect ("open") on November 9, 1965; that triggered a sequence of escalating line overloads that cascaded almost instantaneously down the main trunk lines. Additional lines failed, separating plants from cities that used their power. Generating plants in the New York City area then shut down automatically to prevent overloads to their turbines. The entire Northeast of the United States and large parts of Canada were plunged into an eighteen-hour blackout.

Political process is not governed by the second law, but at its worst, it can be even more disruptive. Regulators historically viewed huge plants and the sprawling grid as "natural monopolies," which meant that any one area could have only one efficient provider of power. That, in turn, meant that prices would have to be regulated. Until recently, that meant regulating both the power plants and the wires, because they had to be engineered and built together, roughly as if each generating plant had its own dedicated grid. Economically separating the two gradually became feasible, however, as the grid developed into a densely interconnected mesh that could perhaps—in principle—pick up and drop off power almost anywhere. But there were more ways to do the separating wrong than to do it right, and wrong ways were tried first.

A first lurch toward regulatory chaos came in the dismal 1970s, with the rise of OPEC and the fall of TMI. The 1978 Public Utilities Regulatory Policies Act required utilities to connect their grids to nonutility windmill farms and wood and trash plants—and to pay premium prices for the tiny trickles of electricity generated by these ridiculously uneconomic and unreliable technologies. In the end, the 1978 "deregulation" forced $40 billion of wasted utility investment in tiny quantities of expensive, unreliable power and grid extensions to bankrupt trash-burning centers and the like.[3] A company by the name of Enron would emerge to pick up a few pieces of this action. Its wind farms generated negligible revenues and endless headaches; Enron had been trying to unload them for some time before the company collapsed in December 2001.

The Energy Policy Act of 1992 created a more significant opportunity, and Enron, to its credit, grabbed it. The new Act really did deregulate something important—the prices that utilities could charge each other for power they shipped across state lines. As expected, this led to a lot more trading of power between utilities, and spawned new brokers of power—with Enron quickly emerging as the most aggressive among them.

The deregulation of interstate power sales had a number of other consequences, some of them good, but most immediately, it increased the incentive to stretch out the grid farther still. Independent power merchants could shake off regulators by taking their kilowatt-hours (kWh) across state lines, so they did. Utilities split themselves up into grid companies, which still had their prices regulated, and generating companies —"merchant power producers"—which no longer did. Before it ever lights a bulb or a computer screen, over half of all our power is now traded, commodity-like, among wholesalers. The prices charged by power traders for their brokerage services aren't significantly regulated, and these traders have taken utilities on as wild a ride as any to be found in the commodities-futures amusement park. Although Enron itself failed spectacularly, many others in the same line of work still prosper.

The deregulation of plants attracted investment in the new, unregulated power plants—merchant generators now own about one-third of the nation's generating capacity, and account for almost all new plant

FIGURE 4.2   U.S. Electricity Transmission

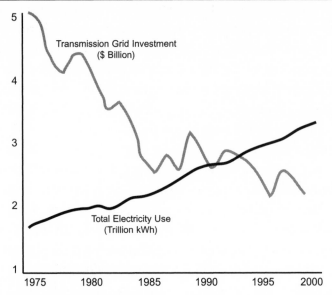

Source: Eric Hirst and Brendan Kirby, "Transmission Planning for a Restructuring U.S. Electricity Industry," Edison Electric Institute, June 2001; EIA, *Annual Energy Review 2003*.

The deregulation of interstate electric power sales promoted investment in new, unregulated power plants, but reduced incentives to invest in the grid.

construction.[4] To stay well clear of regulation, however, they steer well clear of wires—they need grid to get their power to market, but they don't care which market it gets to several state lines to the west or east. As for the grid companies, they remain snarled in regulation, their rates have been kept low, and investment in the wires has plummeted.

Meanwhile, throughout this period consumers were being invited to strain the deteriorating grid to the point of collapse. High-voltage power lines tend to get hot—this is hardly surprising, given that they convey as much power as is produced by the engines of a jumbo jet. When they do get hot, they stretch and sag, and if they sag low enough to touch a tree, a huge wave of power races through the line into the trunk, the tree is incinerated, the surge of power triggers explosives that activate enormous circuit-breakers 50 miles away, and the entire line shuts down. Sagging lines aren't the only problem—from end to end, the grid's switches and transformers can handle only so much power, and any transient surge of power that pushes things over the top precipitates an equally violent stop. The process of halting a jumbo-jet of power in a fraction of a second is not pretty, and after you do, it takes quite a while to get things started again. And here's the joke: it's your cousin Joey, in his one-bedroom walk-up in the Bronx, who—by simply sticking a Pop-Tart into his toaster—can bring it all down.

Joey is ultimately in charge because on a hot summer day, the amount of power moving through the grid gets pushed over the top by the very last toaster that kicks in—that's the one that draws just one watt of power beyond what the gigawatt line can handle 50 miles away. The grid is of course designed so that on most days, and for most hours of every day, there's plenty of capacity to deliver the goods. It's the peak loads that kill you. The straightforward solution would be to price power—and thus use of the grid—much higher at 2:00 P.M. in August than at 4:00 A.M. in December. But instead, utility regulators nationwide still largely insist on flat-rate pricing for residential users. Joey doesn't know he's about to black out New York, because his electric meter hasn't told him.

CALIFORNIA'S SLOUCH INTO NIGHT

It was the convergence of these trends that brought down not just California's grid, but its governor too—the hapless Gray Davis, who happened to be at the helm when the lights went out. It had been in the 1970s, two decades before Davis, that the flat-as-Kansas prognosticators had persuaded California authorities that demand for power was leveling off and would soon decline. The state had all the light bulbs, motors, office equipment, and air conditioning it would need. Efficiency and conservation were going to take over from there on out. And in any event, neighboring states had more capacity than they could use. Oregon, Utah, Arizona, or Montana could breathe the dirty air, while California would import whatever modest amounts of additional power it might need, and fill in any small gaps that might remain with friendly, renewable "micro-generation." California, in short, had already built its last big power plant.

For the next two decades, it was almost impossible to commission a new plant in the Golden State. In 1996, the state's regulators took things a step further, and directed the utilities to sell all their power plants to independent merchants. To make sure utilities couldn't sneak back into the generation market through the back door, regulators strictly forbade them to sign any long-term supply contracts. Utilities were to buy power only in the spot market, and collectively, through a newly created "Independent System Operator." And another good thing: consumer rates were to be cut by 10 percent immediately and price-capped thereafter.

All of this was called "deregulation." And so it was, in the flat-as-Kansas view of things. Too much power would soon be chasing too little demand, prices were bound to fall, so it wasn't really regulation at all to order a cut in retail rates immediately. Spot prices were bound to be lower than long-term contracts, so forbidding the latter wasn't regulation either. The independent power producers could spend their own money as they pleased, so long as they kept things clean in California, which they could easily do by moving out, which they did. Out-of-state systems from British Columbia to Arizona now account for about one-quarter of the peak capacity on which California depends.[5]

All this created a marvelous opportunity for Enron, the pioneer of power arbitrage. Enron had been a real energy company at one point, pumping real gas to real customers, but its genius, while it lasted, was to get out of regulated energy and into deregulated contracts. It moved into commodities trading—gas in 1989, electrons in 1994, and bandwidth in 1999. The profits on the trades—of cubic feet of gas it didn't extract or burn, of kilowatt-hours it didn't generate, and of fiber-optic lines it didn't light—sent Enron's revenues soaring. The company extended its trading operations into pulp and paper, plastics and metals. It ended up writing new financial paper faster than it inspired new trust, the paper grew increasingly speculative, and the trust collapsed. Having staked its power supplies on so much grid, so much resale, and so much Enron,

California went down with it. Other high-growth states didn't make the same mistakes, and didn't suffer the same consequences.

California politicians have since scrambled to sign long-term supply contracts, buy turbines, secure gas supplies, acquire new plants, streamline the process of approving new plant construction, and thus expand supply and push down price. For a time the state seriously considered banning all further utility sales of power plants, and talked of seizing other plants by eminent domain. Having chosen to push the generating capacity it needed into neighboring states, California's leadership now declared, in sonorous, Jimmy-Carteresque terms, that "never again can we allow out-of-state profiteers to hold Californians hostage."

RESTORING ORDER

Enron and Gray Davis did not, however, cause the massive August blackout of the Northeast on August 14, 2003. That one was triggered by relatively routine failures in grid equipment, and while it certainly wasn't the first caused by the grid's own, inherent frailties, it was, by quite a margin, the worst in U.S. history. Perhaps most alarming of all was that the northeastern grid wasn't even particularly stressed that day. There was no great heat wave on the East Coast, the major transmission lines weren't running at full load, and there was generating capacity to spare. The nominal first cause of the blackout was a tree interfering with a major power line.

But one tree falling in the forest should not be able to black out the Northeast five seconds later. And it didn't. The tree had triggered an hour-long series of line failures and plant shutdowns in northern Ohio, near Cleveland—and the implications had simply gone unnoticed and unattended to, because a computer had been switched off and a technician was out to lunch. As the followup investigation concluded, the computer tools used to diagnose the state of the grid supervised by the Midwest Independent Transmission System Operator were "under development and not fully mature" when the fateful tree fell in Ohio.[6] The Operator itself, and its computer systems, had been set up in late 2001, under the supervision of federal regulators, to facilitate the transition to competitive power markets across fifteen states from Texas to Ohio, and up into Manitoba, Canada. On August 14, a grid moving gigawatts (GW) of power collapsed because it failed to move a couple of screens' worth of data and execute a few hundred bits' worth of digital logic.

Electric power is highly ordered from the get-go, but maintaining order across thousands of miles of high-voltage wires is a quite separate and even greater challenge. Because they are so long and carry so much current, the wires store tremendous amounts of power in the electric and magnetic fields that surround them. They have tremendous electrical inertia, and when things change abruptly at one end, the wires themselves act like massive malignant generators that knock voltage and current out of phase and send huge amounts of "reactive power" sloshing up and down the system, like waves in a bathtub.

Grid engineers maintain order, if they can, at "interties" and "substations." These switching points are supposed to isolate problems and flatten out the waves of power by routing power in and out of different lines and through huge transformers and capacitors. High-power switches thus impose order on the grid much as microscopic gates impose logic on a Pentium. The switches are controlled, in turn, by the grid's "supervisory control and data acquisition" (SCADA) networks—which move the bits that control the power. Sensors and dedicated communications links feed information about the state of the grid to regional transmission authorities and utility control centers, and the latter control the switches.

With real-time access to SCADA networks in Ohio, utilities across the Northeast would have seen the August 14 problem coming many minutes, if not hours, before it hit, and could have activated protective switches before the giant wave swept east to overpower them. But utility SCADA networks have evolved piecemeal over the decades, and in the deregulatory scramble of the 1990s, regulators had pushed the physical interconnection of power lines out ahead of the interconnection of data networks. The software systems needed to provide automated monitoring and control across systems had not been deployed. On-site power networks in countless factories and data centers were monitored far more closely, and made much more sophisticated use of predictive failure algorithms.

The grid's key switches hadn't kept pace, either. To this day, almost all the grid's logic is still provided by electromechanical switches—massive, spring-loaded devices that take at least fractions of seconds, an eternity by electrical standards, to actuate. Some years ago, however, ultra-high-power silicon switches reached the point where they could be deployed to control grid power flows much faster, more precisely, and more reliably. These truck-size cabinets contain arrays of solid-state switches that can handle up to 35 MW. They already play key roles in securing power supplies at military bases, airport control hubs, and data and telecom centers. At ultra-high-power levels—up to 100 megawatts—enormous custom-built arrays of solid-state switches can be used to interconnect and isolate high-power transmission lines; but thus far, they've only been deployed at about fifty grid-level interconnection points worldwide.

With advanced control software, interconnected data networks, and high-speed, high-power switches at key locations, the grid could readily be made as smart as it is powerful. Power suppliers know where to put the software and the switches. What regulators entirely failed to give them, however, was any economic incentive to deploy them—the prices suppliers could charge were set too low, with no premium for maintaining a more reliable grid or penalty for failing to do so. However unwittingly, regulators contrived to channel investment capital away from the wires that needed it the most. Had regulators merely left the power plants and wires under common ownership and control, the owners of the plants would have continued to upgrade their own wires to get their own power to market. Or had regulators at least taken the stand-alone grid seriously, they would have allowed the grid owners to charge considerably more for carriage on their wires—while at the same time requiring aggressive new investment in the wires and establishing a framework of penalties and give-backs to kick in when the grid nevertheless failed. But none of this was done.

Nor had anyone made the slightest effort to reconcile the consumer's habits—cousin Joey's—with the realities of the laggard grid. Better flow of information to the end user would not have averted 8/14, but it would have saved California. In terms of *average* daily demand in 2000, even California—albeit with the help of distant neighbors—still had plenty of boilers and turbines to supply electrons, and enough grid to get them where they were needed. The system was strained beyond the point of collapse only during afternoon hours on hot days, when air conditioners kicked in and loads peaked—much as they do during rush hour on the highways of Los Angeles. So, having let itself become far too reliant on distant power plants and the Enron family of power brokers, even California could still have averted its rolling blackouts in 2000 with smarter pricing.

The technology needed was readily at hand to correct the problem. Smart electric meters with two-way communications capabilities built in can now easily replace the antiquated system of rates set out in the fine print of utility bills and a human meter reader pounding the pavement once a month. And with communicative cash registers in place, it becomes easy to set smart prices. Nothing very complicated is needed—just push prices up four hours a day, and down the other twenty. And keep pushing—both ways, on a revenue-neutral basis, lowering off-peak prices while raising on-peak ones—until the congestion abates.

It is often said that such schemes are unfair to the working man and woman, who can't easily change the rhythms of their daily lives. But in these high-tech times, sensible pricing will quickly mobilize a raft of technologies that will improve productivity and the quality of life all around. Cooling, for example, can be stockpiled—for an hour or so with thermostats smart enough to get a jump on the electric rush hour, or around the clock with more elaborate water or ice-based thermal reservoirs. Up to a point, consumers can time-shift electricity itself, using batteries and other storage devices. Dishwashers, washing machines, and other appliances can readily be equipped with timers, to run when rates are low and the grid has capacity to spare. The smart appliance, in short, can be enlisted to create many of the same economies as the grid itself, by ensuring much flatter demand on the wires and the power plants.

Ironically, this is a prescription for more Enrons, though without fake accounting. Large users already sign "interruptible power" contracts with their largest industrial customers—contracts that permit the utility to hand over part of its traditional responsibilities to privately owned, on-premises generators, and place the customer in the position of a power trader, at least as between its own generators and the utility's. Smart meters and time-of-use pricing would give every user similar authority to trade 2:00 P.M. power for 4:00 A.M. power.

Even small nudges in that direction have big payoffs. Just as highway gridlock feeds on itself, peak loads on power lines heat up wires, which raises their resistance, making them heat up still more. Peak-flattening incentives thus have a self-amplifying magic; even when they start out revenue-neutral, they end up paying back premiums—off-peak prices fall more than peak prices rise.

MORE PLANTS AND SHORTER WIRES

In the end, however, there is no escaping the terrible vulnerability of the magnificently efficient grid. Consider the city of New York. Peak demand: about 11 gigawatts. Generation capacity within city limits: 8.8 gigawatts. About 3.7 gigawatts are imported via overhead transmission lines that run down from the north to Westchester, where they transmit to underground cables. These lines bring in nuclear and hydroelectric power from New York State, Connecticut, and Quebec. About one gigawatt comes over three lines from the west, through New Jersey, bearing mainly coal-fired power from the Midwest. All the main lines run above ground, across hundreds of miles of open country. According to a 2002 report by the National Academy of Sciences (not specifically addressing New York), "a coordinated attack on a selected set of key points in the [electrical] system could result in a long-term, multistate blackout. While power might be restored in parts of the region within a matter of days or weeks, acute shortages could mandate rolling blackouts for as long as several years."[7]

There is only one certain answer to both the economic and the physical hazards of too much grid: build more power plants, closer to where the power is needed, and thus shorten the wires that define the problem.

The independent vendors grasped this first. As regulators cut them loose to generate power at any price they could, they turned first to existing coal and nuclear facilities, many of them built in an earlier era when they could be situated close to where their power was actually needed. In the 1990s, improved maintenance and management allowed top-notch nuclear operators to raise the run time of their plants from about 60 percent of the hours in a year to 90 percent. Coal operators pushed their run times from 60 to around 70 percent. And there is still room for more growth. Nuclear plants are being "up-rated" to run 5 to 15 percent hotter, and in 2003 the Bush administration ruled that refurbished coal plants would not be treated as "new sources" of pollution, requiring scrubbers that might otherwise make the refurbishment uneconomical. Overall, another 80 gigawatts or so—about 10 percent of current total U.S. capacity—will be squeezed out of existing coal and nuclear plants in coming years.

But many new plants had to be built too, and without any certainty that there would be enough new grid to deliver their power. When this became clear, as it finally did toward the end of the 1990s, that left only one practical choice. New coal plants take at least five years to build, and nuclear facilities longer still, and both forms of fuel are shunned in states like California, which need the power most. So utilities and the new merchant generators rushed to build new gas-fired plants, which take only six months to two years, and which are opposed least vehemently by green activists because gas is said to be the cleanest and safest among affordable options.

As a result, gas now generates 19 percent of our power (consuming roughly one-third of the country's gas) and accounts for 35 percent of generating capacity—and it runs two to four times as expensive as coal or nuclear. The economically wise course of action is to use gas to cover just the spread between baseload demand and peak demand (typically, about 15 percent) and to provide another 10 percent as a reserve against unexpected outages. Then use other fuels to cover the baseload. That minimizes the combined cost of fuel (high for gas, low for coal) and capital (high for coal, low for gas). But in many parts of the country, the market has swung far past the optimum mix.

The economic implications are now looming. Gas prices are unlikely to come down any time soon. Domestic gas production has remained virtually flat since 1996, and while imports from Canada are up, gas can't easily be imported from any farther afield. Capacity limits in gas pipelines are putting further pricing pressure on gas-fired power. In some markets, for the first time since the dawn of the electrical age, electricity prices are now rising sharply.

At the same time, however, the new, closer-to-home gas plants are providing only modest improvements in grid reliability. The grid's wires are still long, still exposed, and therefore still vulnerable. Worse still, gas-fired plants depend on gas pipelines—which are even longer, and possibly even more vulnerable than, the wires. Gas-fired power plants in the city of New York, for example, are fed by just two groups of pipelines—three pipelines that bring in gas from the Gulf Coast region, and the Iroquois Gas Transmission system that receives western Canadian gas from the Trans-Canada pipeline in Ontario.

What other alternatives might there be? Only two really, and only one of those can supply more than a few hours of backup power. Some 3 to 5 percent of the public grid's capacity is backed up by arrays of batteries (and ancillary electronics) parked under desktops or in office closets or basements, that cushion delicate equipment from electrical blips and supply power during blackouts ranging from minutes to hours. But private and public entities have also deployed 80 gigawatts of on-site generating capacity—about 10 percent of the capacity that lights the grid—to back up (or substitute for) grid power at critical nodes: telephone switches, wireless cell towers, bank computers, E911 operator centers, police communication networks, hospital emergency rooms, air traffic control, street lights, and the electrically actuated valves and pumps that move water, oil, and gas through pipelines.

With very rare exceptions, these backup generators are powered by diesel engines—which is to say, by trucks, stripped of everything but their power plants. And *these* power plants are fueled by an independent, dispersed, and therefore robust network of fuel tanks, tanker trucks, and pumping stations. They have to be—trucks go everywhere, and have to refuel everywhere. The last line of defense for the perilously efficient grid will ultimately have to be found in the parking lot, outside the power plant. The parking-lot engines may be less thermodynamically efficient, but they are smaller, more dispersed, and thus collectively, less vulnerable.

Here yet again, we may well opt to use more energy to get more order—this time around, the form of order we call reliability, or perhaps homeland security.

It is to this creeping convergence of the two largest sectors of energy economy—grid and highway, big power plants and small—that we turn next.

5

FUELING THE SILICON CAR

The first and only automobile ever built that in itself performs all the labor of Electric Self-Starting, Electric Lighting and Ignition. . . . Switch button control for starting, lighting and fuel regulation. Every motion and function controlled from driver's seat.
  —ADVERTISEMENT FOR THE NEW CADILLAC INTER-STATE (1912)[1]

RECALL THE TEN THOUSAND PONTIACS in the parking lot, pedals to the metal, collectively generating about one gigawatt (GW) of kinetic power—as much power as the electric power plant nearby, but half as efficiently. In key respects, the car engine is an anachronism—the technology has more in common with a steam engine than a microprocessor. It "ought to be possible to establish a coordinated global program to accomplish the strategic goal of completely eliminating the internal combustion engine over, say, a twenty-five year period," Al Gore famously declared in his 1992 book, *Earth in the Balance*.* Certain details aside, Gore was right. Sort of.

The internal combustion engine won't be eliminated in our time; to the contrary, in 2017 there will be three to five times as many internal combustion engines worldwide as there were in 1992. But the engine

*P. 326.

itself will be changed beyond recognition; it will change more in the next decade or so than it did in all the years since Nikolaus Otto built the prototype in 1876. Not because an environmentally responsible new president will see to it, still less because the United Nations will take charge of a "coordinated global program," but because power chip technologies have come of age, and because car manufacturers are now in a worldwide race to build new engines around them. The best thing U.S. policy makers can do is step out of the way and let the market find its own way to the extraordinary future that now beckons.

WEIGHT, POWER, AND INERTIA

There's nothing inherently wrong with big, heavy machines. Huge excavators mine coal a lot more efficiently, hundred-car trains haul it better, and gigawatt plants transform more of it into electric power, than any alternative. What *is* objectionable is the space, weight, and energy a big machine consumes just taking care of itself. This is bad almost everywhere, but nowhere worse than in transportation. Most of a car's weight is in the engine and power transmission systems, and in the elaborate controls that interface man and machine. The heaviest things that a car must move are its own engine and fuel tank. Making the engine light and the fuel very dense is therefore essential. But for all that has already been achieved in that regard, car engines and their drive trains remain abominably heavy.

The numbers would be far worse without the internal combustion engine—so much so that we wouldn't have cars at all. No other technology yet comes close to extracting as much affordable, reliable power out of as few pounds as the engine and fuel that so many green pundits despise. Liquid hydrogen and uranium store far more usable energy, pound for pound, but require a much heavier fuel tank (to keep the hydrogen liquid or compressed), or reactor vessel (to extract power fast from uranium). The gas turbines used to propel jets are very difficult to shrink down to car-scale sizes. The external combustion engines used in stationary electric power plants can extract twice as much useful power out of a pound of far cheaper fuel, but the furnaces and condensers are much heavier. Electric batteries fail dismally on the power-per-pound metric.

But to say that the internal combustion engine is the best technology available today is not to say that it is good. As discussed in chapter 3, it is unhappily inefficient, both thermodynamically and because it requires hauling around so much heavy fuel and hardware—pure deadweight in the business of getting a soccer mom and kids to the field. (See Figure 3.3.) Indeed, since humans first began riding horses, and on through modern aircraft, we have continued to use roughly one ton of vehicle to move one human passenger or a couple of hundred pounds of freight. A ten-to-one deadweight-to-payload ratio means that most of everything we sink into transportation, from steel to gasoline to asphalt, is being consumed in unproductive overhead.

Much of the weight of the engine, most of its cost, and all of the logical complexity are located in the peripherals that surround the core array of pistons and cylinders. The engine's "logic" still consists mainly of primitive things that go "click-click," just as they did in James Watts' regulator. Linkages, rocker arms, contacts, and valves flap back and forth, and are typically stopped by colliding with something else. Shafts, belts, pulleys, chains, gears, calipers, throttles, and valves push and pull, open and close, twist and spin in barely adequate synchrony. Elaborately designed arrays of levers, gears, wheels, and cam shafts not only transmit the power but also impose upon it a desired trajectory and timing— just as the insides of an old mechanical watch both power and time the movement of hands around the watch face. A mechanical watch has a certain beauty in its diminutive complexity, but its much bigger sibling under the hood of the car has the look and feel of a nest of thrashing steel snakes. Its engineering aesthetics are altogether ugly.

Still, over the past century, these systems have evolved to incorporate an enormous base of intellectual property and manufacturing know-how—and this ties the industry all the more tightly to the gasoline-fueled core. It is an enormous undertaking to rebuild this entire structure within the framework of a passenger car, an inherently dangerous, pervasively regulated, incessantly litigated, mass-market consumer product. Even if car companies were simply handed a fully functional, noncombustion, hydrogen-fueled, electricity-generating power plant to replace the gasoline-fueled engine that we all rely on today, they couldn't just drop it in under the hood. All the rest of the power train would first have to be redesigned around the new device, and then exhaustively tested and vetted before any manufacturer would dare bring any such hydrogen-electric car to market. Starting down such a road would be a bet-the-company gamble. That just isn't the sort of thing prudent engineers and responsible capitalists do at General Motors, not when the only payoff at the end of the road is the warm approbation of passionate greens.

This, of course, is why those same greens call for a "coordinated global program" and "strategic goal," by which they mean government initiative, government mission, government mandate, government money. But even if a president of that persuasion were to win the White House someday, it is inconceivable that Congress—which can't even muster the votes to add a mile or two to federal fuel-economy standards—would go along. In political circles, as on the road, the internal combustion engine is not only powerful, it has monstrous inertia. It can haul itself down a highway all right, but in Washington, no one can budge it.

Detroit, however, *can* change its engines and does—not through bet-the-company initiatives, but piece by piece. That's how a Model T's engine evolved into a Hummer's. Many, of course, see the Hummer as a compelling argument not to trust the market. But the next twenty years promise spontaneous evolution of a quite different sort. The case for laissez faire today is not that the market knew best when it gave us the internal combustion engine, though it did—it is that the market now knows how to change that engine beyond recognition, and it will.

The immediate opportunity doesn't center on fuel cells or any other scheme to retire the internal combustion engine itself. That might happen ultimately, but not soon. All the rest can change, however, and it will, and that may—just may—set the stage for someday, at the very end of the process, doing away with the combustion as well.

THE SILICON POWER TRAIN

Etienne Lenoir and Nikolaus Otto built the first electro-thermo-mechanical control systems for their spark ignition engines, and Henry Ford incorporated their basic designs in the Model T. The Model T didn't even have a battery; the modern Buick remains a low-power 14 volt electrical platform, used for starting the engine, igniting the fuel, and running power windows and the CD player.

With that said, however, the grid of the modern car is already quite elaborate—over 5,000 feet of copper wire link hundreds of connectors, five dozen fuses, dozens of relays, and fifty to one hundred power chips. Detroit now spends much more on etched silicon than on steel, and the silicon share keeps growing. Electric demand on the automobile platform is now rising about 4 percent per year. The total electric load now runs about 1 kW, with a peak load up to 2 kW. On the other hand, the real power train in the Buick is the 100 kW (peak) / 20 kW (average) mechanical one—the one that begins at the piston rods and ultimately powers the wheels.

It is this gap, between 2 kW and 100 kW, that defines the opportunity for truly radical change.[2] Year by year, the car's power train will now be transformed from mechanical-hydraulic to digital-electric. This is happening now—and it could not have happened a decade ago—because until quite recently, it was too expensive to replace click-click mechanical logic with silicon. But silicon power chips are semiconductors, and just as with silicon logic, the price of silicon-centered power control has been plummeting and performance has been rising as fast. For the first time in history, semiconductors can control kilowatts (kW) of power cheaper, faster, more precisely, and reliably, in far less space, than mechanical alternatives.[3]

For the next decade, at least, it isn't the gasoline, piston, or cylinder that will be displaced, it's all the rest—most of what comes before and after combustion, most of the click-click tangle that transforms the raw explosive force of the piston into a smooth cruise down the highway. The car's mechanical-hydraulic power train will give way to a constellation of servo motors, sensors, silicon electric drivers, and digital power switches.

FIGURE 5.1   Mechanical Drive Train

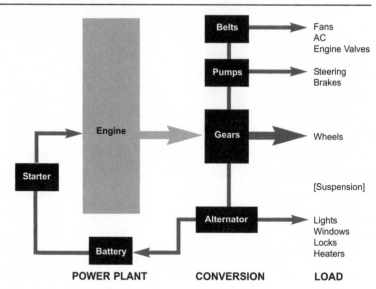

Source: Peter Huber and Mark Mills, "The Powerchip Paradigm II: Broadband Power," *Digital Power Report*, Dec. 2000, www.digitalpowergroup.com.

A car engine's logic still consists mainly of primitive linkages, rocker arms, contacts, valves, and gears. Much of the weight of the conventional car engine, most of its cost, and all of the logical complexity, are located in these peripherals that surround the pistons and cylinders at the core.

Belt-driven radiator cooling fans are already being replaced by silicon-controlled electric cooling. Electric water and oil pumps are displacing belt and pulley driven pumps. Electrohydraulic brakes are already incorporated in high-end cars from Mercedes and BMW. These brakes are more powerful, easier to modulate, and less prone to fade; all-electric brakes will follow. BMWs, Corvettes, and a fast-growing number of less expensive cars have electronic throttles, in which the gas pedal sends electrical instructions to a microprocessor that electronically controls the fuel injection system. High-end automatic transmission systems are controlled by a suite of logic and power chips that take their cues from the driver, the wheels, and an array of engine sensors. Drive-by-wire electric power steering began appearing in some production vehicles in 2001 and will be in almost every car within a decade. Passive, reactive, energy-dissipating springs and shock absorbers will be displaced as well,

FIGURE 5.2   Electric Drive Train

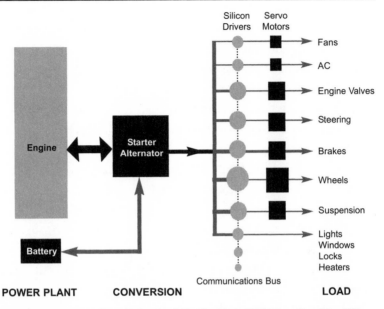

Source: Peter Huber and Mark Mills, "The Powerchip Paradigm II: Broadband Power," *Digital Power Report*, Dec. 2000, www.digitalpowergroup.com.

The car's power train is being transformed from mechanical-hydraulic to digital-electric because low-cost semiconductors can now control kilowatts of power faster, more precisely, more reliably, and in less space, than mechanical alternatives. Car makers already spend more on etched silicon than on steel.

by an active array of powerful linear motors that move wheels vertically as needed to maintain traction beneath and a smooth ride above. Look-ahead technology will discern bumps and holes in the road and move wheels ahead of the (vertical) curve, leading to dramatic improvements in traction, ride comfort, and fuel economy.

James Watt's "regulator," as adapted for internal combustion, is going electric too, with dramatic payoffs in performance. Silicon-controlled electric actuators are now set to displace the steel camshaft on every valved engine. Put each valve under precise, direct, digital-electric control, actuated independently by its own compact electric motor—open and close each valve as dictated by current engine temperature, terrain, load, and countless other variables—and in effect, you continuously retune the engine for peak performance. Belts, shafts, and chains melt away. Everything shrinks, everything gets lighter, and every aspect of performance improves—dramatically.

The last step in this evolution will be the largest: silicon and electric power will knock out the entire gear box, drive shaft, differential, and related hardware—all of which disappear when direct electric drives end up turning the wheels. As noted in chapter 2, power chips now make it possible to build high-power motors the size of a coffee can, and prices are dropping fast.

When such motors finally begin driving the wheels, the entire output of the engine—anywhere from 20 kW to 100 kW as measured now in standard electrical units of power—will have to be converted immediately into electricity before it is distributed, used, or stored throughout the car. It will take heavy-duty wiring and substantial silicon drives and electric motors to propel a hybrid-electric SUV down a highway at 70 mph—but they'll be far smaller than the steel structures in today's power train. Cars will shed many hundreds of pounds, and every key aspect of performance will improve considerably.

A far-fetched scenario? General Electric's 6,000 horsepower diesel-electric AC6000CW locomotive is powered by an enormous diesel-fueled engine-driven generator; everything beyond is electric. Komatsu's 930E—a monster mining truck with 300 ton capacity—is propelled by a 2 megawatt (MW) Detroit diesel-electric generator. Everything else, right down to the 12-foot wheels, is driven electrically. All-electric drives already control fighter jets and submarines. The surface ships now on the Navy's drawing boards are all-electric, from the propeller to the guns. Electric drives are taking over because an electrical bus can convey far more power in much smaller, lighter conduits, and convey it far more precisely and reliably, than even the best designed mechanical drive train.

Likewise in transportation systems frugal enough to be powered by batteries. Much in the news in early 2002, Dean Kamen's "Segway" is an 80-pound, two-wheel scooter-like platform that somehow senses where you want to go and goes there—riding it feels like skiing without the snow. The throttle, brake, and steering wheel are all in the machine's own sensors, completely invisible to the rider, and far more responsive than any conventional set of manual controls. Just lean your body to speed up, slow down, or stop altogether. Kamen's grand hopes notwithstanding, the Segway isn't the transportation platform of the future, and its batteries certainly won't displace gasoline any time soon, but its all-electric, digital-power drive train is indeed the future.

Detroit is starting with luxury cars at the top of the market to go there too, because even when it costs more, the electric power train delivers far better performance all around. Regulatory meddling is more likely to slow down this process than to speed it up, by getting in the way of a fiercely competitive race to build an inherently better power train at the lowest possible cost.

CONVERGENCE

The policy implications are enormous. With a fully electric power train, most of the car—everything but its prime mover, whatever that may end up being—now looks like a giant electrical appliance. Given where battery technology is today, this appliance won't be able to run any great distance on batteries alone, but it will nevertheless have to have a substantial battery pack on-board to provide surges of current when needed. This creates, from the get-go, the possibility of at least some opportunistic "refueling" of the car from the grid. Most of our cars spend most of their time in driveways and parking lots, and a good bit more standing still at traffic lights. As cars grow progressively more electric, the infrastructure for recharging their batteries from the grid will grow apace.

The basic economics strongly favor moving power from grid to car, if you can, and the rise of the plug-in hybrid will have a huge impact on our demand for oil. Burning $2-a-gallon gasoline, the power generated by current hybrid-car engines runs about 35 cents per kilowatt-hour (kWh). Fueled by $40-dollar-per-ton coal, many utilities sell off-peak power for 2 to 4 cents, and the nationwide average residential price is about 8.5 cents. Coal kilowatts are so much cheaper because coal is a low-grade fuel, and because a power plant's huge furnace, boiler, and turbine are a lot more efficient than a V-8. All-electric vehicles flopped in the 1990s because batteries can't store enough power to provide range for long weekend trips. But plug-in hybrids still have the gasoline tank too, and the vast majority of the most fuel-hungry trips are under 6 miles—well within the range of 2 to 5 kilowatt-hour capacity of the on-board nickel-metal-hydride batteries in hybrids already on the road, and easily within the range of emerging automotive-class lithium batteries. The technology for replacing (roughly) one pint of gasoline with one pound of coal to feed one kilowatt-hour of power to the wheels is now here.

Or the power can flow in the opposite direction, and there is reason to do that, too. The hybrid electric cars already on the road—with models from Toyota, Lexus, Honda, GM, and Ford—have 10 to 50 kW power plants, with primary fuel on-board in the gas tank. The hybrid electric trucks and buses now emerging will have 100 to 300 kW. The emergence of this new infrastructure creates the possibility of linking the transportation sector's mobile and highly distributed infrastructure of fuel tanks, engines, and generators—or fuel cells, if they ever fulfill their promise—directly to electrical breaker boxes in residences, small offices, and larger buildings. When it does finally arrive, simple, safe bridging from the transportation sector's power plants to building-level electric grids will offer potentially enormous improvements in the resilience and overall reliability of our power supplies. Indeed, as discussed in the previous chapter, almost all of the backup generators already deployed today are diesel engines—truck engines without the truck.

The convergence of the infrastructure of power in the transportation sector and the grid will transform the energy landscape beyond recognition. As noted earlier, electricity accounts for about 40 percent of our current fuel consumption, transportation another 30 percent. But there is almost no overlap between the electric sector's main fuels—coal, uranium, gas, and water (hydroelectric)—and the transportation sector's. There will be when the car is transformed into an electrical appliance, with a compact 50-kilowatt electric power plant on-board. The fuels and the fuel-conversion technologies will converge. We can then, if we wish, light our personal grid with gasoline or diesel fuel, or propel our cars with uranium or coal. That changes everything.

Power Plants on Wheels

One way or another, an electric power train requires electricity—lots of it, far more than car engines currently generate. If the combustion engine remains the prime mover, it must get hooked up to a larger electric generator, one big enough to convert most—eventually *all*—of the engine's output into electric power.

Nearly all of today's cars still have just the opposite—an electric starter motor that's big enough to turn over the whole engine, but not a generator big enough to dispatch comparable amounts of electricity the other way. The next-generation integrated high-power alternator/starter motors have already been incorporated in BMWs and Benzes, and in Ford and GM trucks; about half of all new cars will have them by 2010. These units will supply the car with abundant, efficiently generated electric power, in a much lighter package, that will provide a virtually instant engine start as well. A 42 volt grid to replace the existing 14 volt grid is the other half of this threshold transformation—lower-voltage wires just can't convey large amounts of power efficiently. A new 42 volt industry standard emerged recently, and half of global automobile production will be on a 42 volt platform within the next decade.

As this process unfolds, the engineering focus will shift inexorably toward finding the most efficient means to generate electricity on-board. Today's diesel-electric trains and monster trucks use big diesel engines to generate electric power, just as factories and office buildings use similar engines for backup electricity. It is the fuel cell, however, that attracts the most attention of visionaries and critics of the internal combustion engine. This is the one technology that might yet prove Al Gore right—not "sort of," but completely.

Among energy technologies that almost no one yet uses, or is likely to use any time soon, this one has certainly attracted the most passionate and sustained interest. Remarkably elegant in its basic operation, the fuel cell—like the solar cell—transforms fuel into electricity in a single step, completely bypassing the furnace, turbine, and generator. The alkali fuel cells used by NASA offer the highest power-to-weight ratio of any electrical generator ever devised, in a technically magnificent—though tricky and dangerous—package. What has attracted recent interest is a much safer cell design that runs relatively cool (at about 93°C) and depends on a proton exchange membrane (PEM) impregnated with a platinum catalyst to promote a hydrogen-oxygen reaction.

The surge of interest in fuel cells was triggered by breakthroughs in platinum chemistry and solid electrolytes achieved only a decade or so ago. The first PEMs were built in 1953, but it wasn't until recently that economical methods were found to deposit ten-atom-sized platinum particles on pure carbon that is bound to a semipermeable Teflon-like "Nafion" membrane, a tenth of an inch thick. This architecture sharply reduces the amount of catalyst, and thus the cost of the membrane and the hydrogen-electricity conversions.

Because they consume pure hydrogen and maintain low temperatures,* fuel cells emit nothing but water, so they can be located not only in cars—which then run ultra-clean—but also in any home, office, or factory, as close at hand as a dishwasher or a desktop computer, thus directly substituting for power from the grid.

The hydrogen, however, is the tricky part. PEMs are quickly poisoned by carbon, so PEM fuel cells can't run directly on hydrocarbon fuels; they require exceptionally pure hydrogen. As we discuss further in chapter 7, hydrogen can be extracted from natural gas, and can then be viewed as a purified, carbon-free *fuel,* but this requires a lot of hydrocarbon—far more than just burning both the hydrogen and the carbon of the "hydrocarbon" together.

Alternatively, hydrogen can be extracted from water: here the hydrogen really serves the same function as a rechargeable battery. Not a very efficient one—it takes about 4 kilowatt-hours of electricity pumped into the water to get one kilowatt-hour of electricity back out—but such losses may be acceptable if very cheap sources of electricity are at hand,

*The "low temperature" of the fuel cell reflects the elegance and complexity of catalysis on the PEM. As discussed in chapters 3 and 4, all normal thermal engines depend on temperature *difference* to extract useful work from burning fuel. But the right catalyst can be structured to "burn" hydrogen and oxygen atom by atom, on the surface of the catalyst itself, to produce not heat but something even "hotter" (in thermodynamic terms)—a flow of pure electrons.

and the main objective is to synthesize high-grade energy in a form suitable for powering fuel cells that can't be connected to a gas line. NASA, for example, has plans to use a bidirectional fuel cell to make hydrogen from solar power on the bright side of an orbit, and to generate electricity on the dark side. With an affordable PEM at hand to provide the hydrogen-to-electricity interface, a tank of pure hydrogen is the closest practical thing to a tank of stored electrons.

Back on Earth, the distribution of hydrogen presents an additional challenge—the gas is difficult to liquify, it's viscous when pumped under high pressure, and it sneaks through even the tiniest cracks, which makes for serious safety problems. For smaller, portable applications, however, hydrogen can be adsorbed onto metal hydrides. Other storage media will undoubtedly be developed, too. Sodium borohydride shows early promise, for example. With some suitable technology or chemistry to store hydrogen on-board, fuel cells could eventually replace the combustion engine that recharges the nickel-metal-hydride battery pack in hybrid-electric cars.

This then sets the stage for a visionary energy future even rosier than the most determined enemies of internal combustion have hoped for—a convergence of power plants, with the one parked in the driveway hooking up to the breaker box in the basement, to supply the home with power whenever the car isn't out on the road. If solar-electric power is used to extract hydrogen from water, there will be no more need for big central power plants and filling stations. We will decommission the nuclear power plants, shut down the strip mines, and scrap the offshore oil wells. Silicon and hydrogen will completely displace uranium and carbon. We will be all-solar and all-electric. Silicon, wind, and sun—earth, air, and fire—will move the fourth element, water, to power every aspect of our lives.

URANIUM IN THE GAS TANK

While visionaries map out a solar-hydrogen future, established utilities and their customers respond on shorter time frames with more practical and concrete choices. Electricity can indeed be stored as hydrogen extracted from water. But today, millions of digital enterprises find it more practical to store it in a witches' brew of sulfuric acid and lead.

When current is drawn from a lead-acid battery, lead and lead oxide react with sulfuric acid to produce lead sulfate and water; the process is reversed when the battery is recharged.* Portable devices use lithium, cadmium, nickel, silver, and zinc in their batteries—these elements offer much higher power densities than lead and sulfur, but at much higher cost. In most stationary applications, the best balance between cost and power density is still lead, the ancient metal responsible for poisoning the wine (and thus accelerating the decline) of the Roman Empire.

Lead-acid batteries thus provide storage in tens of millions of uninterruptible power supplies for desktop computers, office networks, and telephone central offices. Utilities have explored the possibilities, too, on a larger scale. In the mid-1980s, Southern California Edison installed and operated 8,256 telecom-type lead-acid batteries in a massive 10 megawatt array, 50 miles outside of Los Angeles. The idea was to test out the economics of time-shifting electricity in much the same way as a TiVo time-shifts a television show. Other chemicals have been tried, too. Regenesys, a subsidiary of Britain's National Power utility, is developing a hectare-sized sodium bromide/sodium polysulfide reversible fuel cell to serve as a 5 to 500 MW regenerative system that will consume power off-peak and produce on-peak.

When more than eight thousand batteries are deployed to store power, however, they don't all have to be piled up in one place. Placed in basements, they can interface with solar panels or windmills, just as proponents of renewable fuels hope they will be. Or—perhaps even more economically—they can simply rely on the grid itself. The most hassle-free substitute for grid power at 2:00 P.M. is grid power delivered to a

---

*Water in the battery is also electrolyzed into hydrogen and oxygen in a side reaction; the breakthrough in the development of sealed batteries was a silica catalyst that promotes the recombination of the dangerous hydrogen-oxygen within the cell. The fuel cell may be the battery of the future, but hydrogen-oxygen chemistry remains just a dangerous nuisance in the batteries of the present.

lead-acid cell twelve hours earlier. And if priced at its 2-cent marginal cost—which it should be, as we discussed in chapter 4—off-peak grid power certainly remains the cheapest option, too. Sunlight and wind may be free, but solar panels and windmills are expensive. The most cost-effective way to charge up the lead battery leads straight back to coal.

A future of lead, sulfur, and carbon is not exactly the clean-energy future that most environmentalists have in mind, and regulatory obstacles could easily derail it. The important point, however, is that such options multiply faster than policy planners and environmental activists often recognize. The fuel cell is attractive only if the best way to store electricity is in hydrogen and the cheapest way to generate electricity is with a solar cell. By all current indications, however, lead and coal will prove to be a much more economical pair.

Or perhaps hydrogen and uranium. Even if hydrogen beats lead in the storing of electricity, solar cells could well lose out to nuclear fuel in generating the electricity needed to extract the hydrogen from water. Uranium and sunlight are both thin, highly dispersed fuels, but we know how to enrich the uranium very effectively, after which it becomes a very compact and—absent misadventure—clean source of power. Ironically, the PEM was originally developed for the Navy, in 1973, for replenishing air supplies on nuclear submarines, where electricity is abundant and hydrogen is a dangerous by-product, but oxygen is scarce. If the fuel cell now leads us to the hydrogen economy, the hydrogen could very well lead us back to the nuclear power plant, which runs most economically when it churns out steady gigawatts of power around the clock.

BLIND VISIONARIES

It is impossible to predict the future, but it is possible to predict the future that has already happened, and the electrification of the drive train is one such future. It has already happened in trains, trucks, and subs; it is already well underway in cars. It's happening because the core technologies—power chips—are improving as fast as everything else the semiconductor industry churns out, and because they deliver enormous improvement in performance. And without doubt, this future—this convergence of the electric and transportation sectors of our energy economy—does have enormous implications for energy policy.

But all the rest—whether we will end up powering more of the grid with gasoline, or propelling cars with grid-generated hydrogen, or whether demand for hydrogen will impel massive new deployment of solar cells—all the rest is anyone's guess. And because we can't begin to predict how the economics of these very different alternatives will shake out over the next decade or two, it's a waste of time to try to formulate long-term policies centered on one vision rather than another. For the next decade at least, policy makers with an eye on transportation technology should have the wisdom and courage to stand aside and let the future unfold without them.

6

BULBS, RADIOS,
AND NEGAWATTS

People stood overwhelmed with awe, as if in the presence of the super-
natural. The strange, weird light, exceeded in power only by the sun, rendered the square as light as midday. Men fell on their knees, groans were
uttered at the sight and many were dumb with amazement. *We contem-
plated the new wonder in science as lightning brought down from the heavens.*
—EYEWITNESS ACCOUNT REPORTING THE FIRST DEMONSTRATION
OF ELECTRIC LIGHTING IN WABASH, INDIANA (1880)[1]

A CENTURY AFTER the miracle of Wabash, the awe had given way to
contempt. The only thing that amazed some people about light bulbs in
the 1980s was how wasteful they still were. And what mystified these
pundits was why consumers just wouldn't snap up super-efficient bulbs
that were there for the taking. To be sure, these bulbs cost $20 or more,
but they would certainly pay for themselves over time, in energy saved.

Even sophisticated businesses—run by people who were supposed to
understand elementary economics—seemed to shun them. An analysis
by Lawrence Berkeley Labs painstakingly demonstrated that the longer
expected lifetime and lower electricity use of high-efficiency lighting
"ballasts" (the electronic components that feed fluorescent lights) more
than offset the higher initial cost. Countless companies would earn far

higher rates of return investing in these ballasts—37 to 199 percent, this analysis concluded—than they could through most other uses of their capital. Nevertheless, companies just couldn't bring themselves to buy the high-efficiency ballasts, not until federal and state governments effectively mandated their use in the 1990s.[2]

The peddlers of fluorescent bulbs came to the fore at the end of the dismal 1970s, the convulsive decade of "energy crisis" in America. They peddled "negawatts"—new efficiency as the alternative to new power. Forcing a utility to pay for the super-efficient bulbs deployed by its customers would be cheaper, cleaner, and more efficient than building a new power plant; instead of investing in concrete and coal, the utility would invest in better ballasts and such, which customers would effectively rent from the utility, through their monthly electric bills, whether or not they actually bought the now highly subsidized bulbs. The power stakes were high. Some 10 to 15 percent of a typical household's electricity is used for lighting; this could be cut to well under 5 percent, and the challenge was simply to get the right bulbs into the sockets.

Step by step, the negawatt crowd converted utility regulators and federal energy officials to their cause. Utilities were required to offer coupons and cash rebates to residential consumers for the purchase of fluorescent bulbs. The federal government directed all its building managers to buy nothing else, and kicked in federal dollars to help subsidize household purchases as well. In March 2001, even as it hovered on the verge of bankruptcy, and after a year of rolling blackouts precipitated by its inability to keep power moving through its own grid, California's Pacific Gas & Electric was still giving away free bulbs.

The efficiency debate spans not only bulbs but also refrigerators, washing machines, and, of course, car engines, and we will return to the broader subject in the next chapter. But much can be learned about the dynamics of efficiency by focusing on just one well-defined sector, and lighting is an excellent place to start. What lighting teaches is this: the nabobs of negawatts were wrong—wrong in their economics, and even more wrong about where the technology of photon power was headed. Through the subsidies and regulations that they successfully promoted,

FIGURE 6.1 The Cost of Illumination

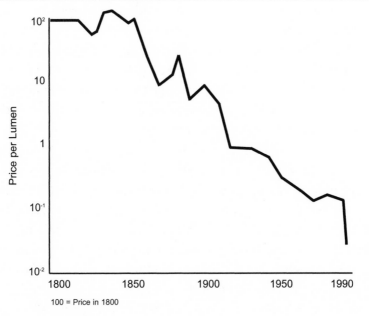

100 = Price in 1800

Source: *Fortune*, Nov. 22, 1999.

The cost of light has dropped ten-thousand-fold over the last two hundred years.

they wasted huge amounts of our money. We would be richer today—and more energy-efficient too—if they had never arrived on the scene at all. Efficiency notwithstanding, we would be using more energy for "lighting" too—a paradox so repellant to the accepted wisdom, so impossibly counterintuitive, that pundits of almost every political stripe simply reject it out of hand. But the facts are what they are. The pundits were guilty of a massive failure of imagination.

EDISON AND DEFOREST

In 1879, the remarkable thing about Edison's new light bulb wasn't that it used a lot of electricity—which it certainly did—but that all that power didn't burn down the house.* Lots of people already knew that a current could be used to heat a wire, but nobody had worked out a practical way to get a wire hot enough to emit much useful light without bursting into flames. Edison solved the problem by mounting a carbonized cotton thread in an evacuated glass bulb. His bulbs soon proved to be cooler, safer, and more practical for most applications than the brush arc lamps that were being used some years earlier, to the astonishment of the good citizens of Wabash, Indiana, among others.

Half a century later, the man who pioneered aerial photography for the Air Force would use a magnesium powder bomb for his flash lamp. Timing was tricky—the camera's shutter had to be precisely synchronized with the release and ignition of a very dangerous incendiary. In 1939, the military had asked MIT's Harold Edgerton to adapt his recently invented stroboscope for nighttime aerial photography. Before the year was out, Edgerton tested a prototype in a B-18 bomber over Boston. In 1943 he delivered specifications for a 43,200 watt-second device capable of taking photographs of the surface 20,000 feet below.†

But many other "bulbs" had appeared on the scene long before Edgerton pushed Edison's technology to such inflammatory extremes. Light is electromagnetic radiation, but so are radio waves; the top end of a cell phone is a bulb of sorts, casting very (very) "red" light (now called "microwaves") outward in all directions, to be discerned a mile or two away by the gimlet-eyed receiving antenna of a cellular base station. We make this point at the outset not to be clever, but because this, in a nutshell, is why "efficiency" policies invariably fail: the more energy-efficient a technology grows, the faster it metastasizes and finds new ap-

---

*An electrical fire did seriously damage the library at one of Edison's first installations—J. P. Morgan's elegant Madison Avenue brownstone in New York. See Jill Jonnes, *Empires of Light: Edison, Tesla, Westinghouse, and the Race to Electrify the World* (Random House, 2003).

†Today, we remember Harold Edgerton for his remarkable photographs of a bullet frozen in flight in the instant after it has burst through an apple, or the beautiful crown formed by a drop of milk in the instant after it collides with a plate.

plications. It is instructive to explore this process just with "light" only because electromagnetic radiation is such a discrete, well-defined form of power.

Edison's bulb had come first, but it turned out to be only one of two path-breaking technologies of the first century of photon power—Lee DeForest's vacuum tube was the second. Both depended on moving electric currents through wires mounted in evacuated glass bulbs. Both radiated photons—Edison's filament as visible light, DeForest's tube as radio waves, by way of Marconi's radio, which had been invented a decade earlier.

Edison's bulb was a very much more convenient producer of light than a gas lamp, and the grid that he built to power his bulbs provided a clean, compact way to deliver power to where the light was needed. DeForest would add two more filaments to Edison's bulb to create the first electronic amplifier. His three-filament tube made possible much more compact and powerful radio transmitters, and more sensitive receivers. This was the technology that made Marconi's radio functional, powerful, and cheap enough to blanket the continent with commercial broadcasts—and to propel a receiver into the parlor of every home.

Thus between Edison in 1879 and DeForest in 1906, the technological stage was set for the next half-century of photon power. That century of lighting revolved, one might say, around a new generation of "gas lamps"—evacuated bulbs of one kind or another, in which electrically heated filaments emitted energetic streams of photons or electrons. The technologies were remarkable advances over what they replaced. Yet both still generated far more heat than useful streams of electromagnetic energy.

When radio waves were pushed into radar bands during World War II, the key technology turned out to be the "cavity magnetron," yet another bulb-like structure, this one ingeniously designed to emit photons at microwave frequencies. Small enough to be held in one hand, and producing 50 kilowatt (kW) one-microsecond pulses of 3000 kilohertz (kHz) radiation 400 times per second, it could light up a large ship at 100 miles, a conning tower at 20 miles, and a periscope at 5 miles. It could also render visible city streets from an aircraft flying above the clouds. The design perfected from the British prototype depended on an 18,000-volt power supply built around a 30-pound magnet and a hockey-puck-sized metal and glass tube. But only one-fourth of the electric power pumped into the magnetron emerged as microwave photons; the rest, once again, emerged as heat.

To this day, photon power remains anchored in the Edison-DeForest legacies. Americans buy some 4 billion incandescent bulbs a year, along with tens of millions of tube-based television and computer displays. X-ray machines depend on massive tubes and 50-kilowatt power supplies to accelerate an electron beam down a 100,000-volt hill and slam it into a tungsten target. High-power radio and television broadcast stations are packed with water-cooled vacuum-tube amplifiers. To light up the antennas perched 500 to 1,000 feet above the ground on a single, typical mast, hundreds of kilowatts of power—as much power as could be delivered by the engines in a small fleet of Buicks—pour through banks of monster 30 kilowatt (or larger) vacuum tubes mounted in metal racks down at the base. And all this just to dispatch information, the very lightest of payloads.

The apogee of Edison's technology, one might say, was the laser—the best, the brightest, and, in key respects, the most ridiculous—descendant of Edison's genius. The theory was brilliant enough to earn Charles H. Townes a Nobel Prize in physics; the design and construction of the first functioning device was an engineering triumph.* Large arrays of studio flash lamps mounted around a ruby crystal were fired frantically to supply the laser with the low-grade light that it used as its input fuel. To produce more laser power, you bought more flash lamps. Thus, behind the magical beam, everything was headed in the wrong direction—toward preposterous inefficiency and mountainous cooling systems.†

---

*The first was actually a maser, emitting microwaves in Marconi's domain, rather than Edison's; Townes then teamed up with his brother-in-law Arthur Schawlow to build the first laser.
†The Department of Energy's NOVA laser for fusion research, with its awesome 100 TW pulses, occupied a three-story building the size of a football field.

To lump radios, X-ray machines, and lasers in with the gas lamps of Victorian England is a bit harsh. But all of these devices reek of the Old World and barely hint at the New. All of them depend on the chaotic excitation of a metal filament or gas in a bottle, or some solid equivalent like a large ruby crystal. All are dreadfully wasteful. Electricity pours in. Heat, mainly—destructive, useless, wasteful heat—pours out.

Getting flames, electric wires, and incendiary bombs to emit light isn't difficult—the challenge has always been to get more light and less heat. The trick is to produce well-ordered streams of photons in narrower frequency bands situated (in frequency terms) both above and below the ambient thermal chaos. And that has proved to be one of the most enduring challenges in the entire realm of power. Along the way—as a mere footnote to the real action—this challenge has also created abundant opportunity for officious peddlers of negawatts. Be that as it may, this much is clear: invent a better light bulb, and it's bound to have quite an impact. It won't lower energy consumption. It will increase it.

BULBS ON A CHIP

The demise of the gas lamp did not begin with fluorescent lights—those were already being developed in the 1890s, almost a century before the negawatt salesmen arrived to tout them so vigorously. It began with the invention of the transistor, in 1948. William Shockley's Bell Labs team had been directed to develop a compact, efficient replacement for De-Forest's vacuum-tube amplifier, and the team delivered. The early transistor radios were the first commercial products to take advantage of the new, solid-state amplifiers. In 1951, Lee DeForest wrote a testy letter to the editor of *Scientific American* in which he insisted that "the general application of the transistor in radio and television receivers is far in the future." He was wrong. Solid-state amplifiers quickly seized an important share of the DeForest territory.

The transistor was ready to amplify weak signals received by a radio receiver almost from the get-go. It would take quite a bit longer, however, to bulk up the transistor to the point where it could handle the higher currents required for *transmitters*. The cell phone revolution, for example, can be traced to Raytheon's development of the first gallium-arsenide (GaAs) monolithic microwave integrated circuit (MMIC) in 1983. GaAs is a remarkable (albeit expensive) semiconductor that can handle very high frequencies. In 1992, Motorola invented a new silicon-chip architecture,* which is now rapidly emerging as dominant for higher-power applications such as broadband wireless, commercial broadcasting, and longer wavelength radar. In 1997, IBM introduced a silicon-germanium radio-frequency chip able to take (low-power) performance up to 75 gigahertz (GHz). To push solid-state amplifiers up above 200 gigahertz, TRW introduced indium-phosphide devices suitable for use in both telecommunications and low-power radar. On the very near horizon are high-power, high-frequency radio transistors based on gallium nitride. The cavity magnetron of World War II can now be replicated in a thumbnail's worth of this enormously challenging man-made crystal.

Solid-state light bulbs took longer. The first one arrived in 1962, when Nick Holonyak, a General Electric researcher, managed to transform one of the three streams of current that flow through a transistor into a stream of light. In certain semiconductors—the exotic III-V semiconductors, so named for the positions their constituent elements occupy in the Periodic Table—the quantum changes in electron states at a semiconductor junction are very efficient at emitting photons. The junction itself, in other words, now acts as a white-hot filament, except that it's neither white nor particularly hot. If the geometry is right, the photons emerge from the surface or side of the junction, and you have a light emitting diode (LED). The color of the light depends on the materials used. Aluminum gallium arsenide shines red, gallium phosphide is green, gallium phosphide with arsenic is yellow, and gallium nitride is blue.

The shift from Edison's filament to quantum technology radically improves performance. Per unit of area and of energy used, semiconductor

---

*It is called the laterally diffused metal-oxide semiconductor (LDMOS).

FIGURE 6.2   The Materials of Illumination

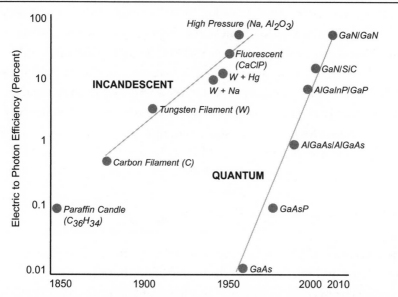

Semiconductor junctions exploit the bizarre phenomena of quantum physics to convert electricity into light far more efficiently than is possible with conventional incandescent technologies. Aluminum gallium arsenide (AlGaAs) shines red, gallium phosphide (GaP) green, gallium phosphide with arsenic (GaAsP) yellow, gallium nitride (GaN) blue.

"filaments" are far brighter than Edison's, which means they can be much more compact, efficient, and cool. In 1987, a company called Cree began developing silicon carbide substrates for blue LEDs; the company recently built a device with a stunning 28 percent electricity-to-photon conversion efficiency—nearly ten times as efficient as typical incandescent bulbs. With so much less junk heat, you don't need an evacuated bulb to keep things from bursting into flames, and the "bulb" shrinks from the size of a pear to the size of a poppy seed. The seed is far less delicate, far brighter per unit area, and can be mounted on just about anything that can supply it with electricity.

Thus, solid-state LEDs have already displaced the old bulbs wherever it is important to supply more light with less power—in battery-powered devices of every description, from wristwatches to emergency exit signs to traffic lights, in portable electronics of all kinds, and in cars, from the dashboard to the tail lights, and soon the headlamps too. Full-color LED displays are possible now that blue LEDs have caught up with the reds and greens. Baseball parks are now erecting huge ones for instant replays. Some 18 million LEDs light the NASDAQ's giant display in New York's Times Square. Over the next couple of decades, solid-state lighting will supersede Edison's filaments almost everywhere else. Even the evacuated tubes in television and computer displays are now rapidly giving way to thin-film transistors, liquid crystals, and, most recently, the brilliant colors of plasma arrays and organic LEDs.

LIGHTING THE WHOLE RAINBOW

The upshot of all this is that we now have unimaginably efficient light bulbs that can shine not just in the visible bands but across broad swaths of the electromagnetic spectrum. So they shine not only in the baseball parks but also (in the radio bands) from every cell phone in the pocket of every fan in the park. And through fiber-optic fibers running beneath it, where Marconi–DeForest radio transmitter now gives way to a reincarnation of Edison's—the laser. And down into the gall bladder during keyhole surgery.

With inexpensive MMIC chips in hand to amplify currents to sufficiently high frequencies, radar engineers can now project radio waves in various millimeter bands that cut through clutter—kilometers of messy air, for example—but that reflect off many things that we really do want to see, such as hard plastic and metal. Millimeter-wave emitters are now used to measure liquid levels in tanks, in chemical, pharmaceutical, and power plants, oil refineries, and countless other industrial settings. They detect voids and deterioration in concrete, pavements, bridges, and railroad beds. They map oil spills, buried hazardous wastes, underground containers, pipes, tunnels, buried mines, ice thickness, and archaeological sites. Space-based systems play major roles in remote sensing of weather and earth resources.

Millimeter-wave radar will soon become a standard feature on cars; it will monitor blind spots to assist with lane changes and parking, alert the driver to collision threats, and preactivate airbags moments before a crash.* Stop-and-go cruise control takes things the obvious next step for city driving. "Cooperative adaptive cruise control" is the inevitable end point—the sensors in cars on the highway spontaneously form their own "local area" networks to coordinate movement with each other, and eventually with an intelligent roadway. Some of these systems are already widely deployed in buses and trucks; they will likely become mandatory for all new commercial vehicles before long. They are being mounted in snowplows to make sure they plow snow, not parked cars or fire hydrants. Comparable systems play a key role in fully autonomous landing systems for aircraft.

The transmitters and receivers of photon power have, in short, now reached the point where it is possible to illuminate, and thus see, almost anything anywhere. Multiple beams are now projected across a huge range of frequencies, most of them invisible to the human eye; this "multispectral imaging" gathers vastly more information than can be revealed by any ordinary light bulb. Some frequencies bounce off surfaces; others punch through obstacles. More penetrating frequencies can provide a chemical fingerprint of the target; less penetrating frequencies discern shape and volume. Such technologies now spread multispectral radiation across the factory floor, the highway, and the battlefield. A decade ago, no one would have imagined such capabilities coming to the mass market. Today, they are not only inevitable but imminent.

All of which is indeed cause for celebration. Except perhaps by those who imagined that fluorescent bulbs might put a real dent in how much electricity we use for "lighting." The new solid-state lights certainly do save electricity—in one narrow band of the electromagnetic spectrum. But almost simultaneously, they give us the means to radiate power across the entire rainbow. The power consumed by all the new, ultra-efficient bulbs far exceeds the power consumed by the old, wasteful ones.

---

*A 77 GHz active cruise control system in the Mercedes-Benz S-class, for example, already slows the vehicle down when it pulls too close to one in front.

FIGURE 6.3   The Spectrum of Photon Power

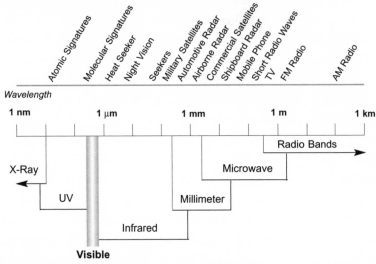

Most wavelengths of "light" are invisible to the human eye. Different wavelengths illuminate, penetrate, and reflect in quite different ways. The new semiconductor "light bulbs" can be configured to emit photons all across the electromagnetic rainbow. The new transmitters and receivers of photon power can thus illuminate and see far more than the unassisted eye.

CONVERGENCE

The new light is so bright, so efficient, and so readily controlled that it is now being used deliberately to produce heat—the original enemy. This is a very important development, because it is impelling a second great convergence in our energy economy.

Recall (from chapter 1) that roughly 40 percent of our raw fuel is currently used to generate electricity, 30 percent for transportation, and 30 percent for heat. The purely thermal sector includes most forms of home heating, along with industrial ovens, welders, chemical processing, and things of that sort. The two principal fuels used in this sector are oil and natural gas—both of which are relatively expensive and in tight supply. It has always been possible to heat directly with electricity, but until now it has rarely been economical. Electricity is such a pure and expensive form of power that it generally doesn't make sense to run it through a coil or filament to produce ordinary heat.

But running it through a solid-state laser is quite another matter. In terms of coherence, and thus power density, and thus overall energetic order, a laser is millions of times better than an incandescent wire. The laser emits in a narrow frequency band, with the waves in phase. By increasing the order, it punches more power through less space—power densities of 20 megawatts per square centimeter (MW/cm^2) are now routine. Soon after the laser was invented, one wag proposed to calibrate a pulsed laser beam's intensity by how many stacked razor blades it could punch through, with the new unit of intensity dubbed a "Gillette." The only other contender for the power-density crown is a utility's high-voltage power line, where power densities can run up to 100 MW/cm^2.

So long as they remained tied to gas-lamp engineering, lasers were, by and large, a brilliant invention in search of a worthwhile application. Industrial uses did slowly emerge—deep ultraviolet excimer lasers are used in the high-resolution photolithography that keeps packing more logic gates onto less silicon wafer, for example. But though they defined a steadily growing market, nobody was going to confuse gas-lamp lasers with, say, the silicon chips that they helped fabricate. By comparison with most datacom- and telecom-equipment markets, lasers were going nowhere fast.

In 1963, however, just a year after Holonyak built the first semiconductor LED, Herbert Kroemer and Zhores Alferov proposed a (Nobel Prize-winning) theory for heterostructure semiconductor devices, in which atomic-scale layers of aluminum and gallium compounds would sandwich carriers and photons to produce coherent laser beams. Other laser-diode architectures emerged too, but it took seven years of additional engineering work by Alferov and Bell Labs to build the first heterostructure device that would become the forerunner of all of today's solid-state lasers.

Semiconductor lasers progressed from aluminum gallium arsenide to indium gallium arsenide phosphide. Power levels began to rise sharply when new semiconductors and new substrates provided more uniform and therefore better light-emitting lattices, better electrical conduits for moving power in, and better thermal conduits for moving heat out. Ten years ago, green diodes offered the highest frequencies available from semiconductor-based devices, with peak powers of about 10 milliwatts; 10-watt devices are common today. Vendors are now pushing deep-ultraviolet devices up the same power curve, into the space currently served by excimer gas lasers. Today, the typical semiconductor laser is a tiny brick-shaped device a few tenths of a millimeter long. Recently developed micro-lasers are just a few thousandths of a millimeter across.

And lasers can be used to pump lasers, pushing power levels higher and higher. The diode-pumped solid-state laser starts with a laser diode that converts over half its electron power into photons, and uses those photons to pump a second-stage laser with photon-to-photon conversion efficiencies of 30 to 50 percent. The overall plug-to-beam efficiency jumps immediately from under one percent to 18 percent or more. The laser is now solid-state from end to end. Every component of the device is now small, and can get a lot smaller. The high-power laser is now set to evolve in the same direction as the high-power computer—shrinking inexorably and marching relentlessly toward low-cost mass production.

A state-of-the art blue laser, for example, now consumes 100 watts of plug power and is the size of a couple of decks of cards; it replaces a 1,500 watts argon gas laser that's ten times as big. Matchbox-sized devices are coming soon. Multiple laser-diode bars arrayed to pump light into a single-crystal rod can readily push device power into kilowatt ranges and beyond. Pulse power can be pushed into the terawatt range.

At these power levels, lasers don't move bits, they move atoms. They fuse powdered metals into finished parts, without any machining, cutting, or joining. They supply ultra-fine heating, soldering, drilling, cutting, and materials processing, with fantastic improvements in speed, precision, and efficiency. They create thermal pulses that can blast metals and other materials off a source and deposit them on a target to create entire new classes of material coatings. They move ink in printers—not just desktop devices but also the mammoth machines used to produce newspapers. They solder optoelectronic chips without destroying the silicon real estate around them, and they supply unequaled precision in the bulk processing of workaday materials—heat treating, welding, polymer bonding, sintering, soldering, epoxy curing, and the hardening, abrading, and milling of surfaces. They are directed at the human body itself to remove hair, reshape the surface of the eye, and cauterize tissue through an endoscope. They dump power into a photoreactive dye that accumulates in cancerous tissues without killing healthy cells nearby.[3] In short, high-power coherent light is now destined to become as important in our energy economy as the automobile.*

EFFICIENCY

Even on their own terms, the negawatt pundits got much of their basic economics just plain wrong.[4] The economics of efficiency depend, of course, on the cost of what efficiency saves. The Lawrence Berkeley Labs study of the economics of ballasts, noted at the beginning of this chapter, started out with the average cost of electricity in 1993 and assumed that the price would remain constant. But the average price of electricity had fallen throughout the preceding century, and it has continued to fall in all but the most clearly mismanaged markets (such as California) since then. Efficiency savings depend, moreover, on the *marginal* cost of electricity—the cost of each additional kilowatt—not the average used in the study. The average includes the cost of the grid—which gets charged to end users, as it must, however little or much power they use—and more efficient bulbs have no near-term impact on that at all. The study also fudged its analysis by excluding businesses that used very little electricity—the businesses that have the least to gain by deploying expensive efficiency.

Finally, the study assumed that the useful life of bulbs is determined by how long they last before they burn out. This was wrong on two counts. For the managers of commercial buildings, much of the cost of

---

*Eamonn Fingleton is one of very few who have noted this important trend in his book *In Praise of Hard Industries* (Buttonwood, 1999). "Next to computer chips, the most important enabling components of the next electronic revolution are undoubtedly lasers and other high-tech light-emitting devices. In contrast with computer chips, however, the highly advanced and important business of manufacturing laser devices gets little attention in the media."

replacing a bulb isn't in the bulb but in the labor. So building managers often replace all the bulbs on a fixed schedule, or whenever the premises are renovated. More to the point, fluorescent lights themselves have improved very rapidly—they have grown far smaller, prices have dropped dramatically, and the garish "Night of the Living Dead" illumination supplied by the bulbs of a decade or two ago has given way to appealing, full-spectrum light. Still more to the point, fluorescent lights themselves are now fast becoming obsolete—overtaken by solid-state light, which is dramatically more efficient and will soon be cheaper, more compact, and superior in every possible respect.

In other words, those who invested heavily in the "efficient" technology of the 1990s wasted their money. They would have done far better—and achieved far more efficiency—by just stashing away the cash for a few years until the next great thing in efficiency materialized. The negawatt crowd ought to be delighted that it has indeed emerged, and perhaps they are. The new light bulbs don't just run much more efficiently than the old bulbs, they can radically improve the energy efficiency of countless forms of industrial heating, welding, and chemical processing, which together account for about 10 percent of the raw energy we consume. And they are steadily encroaching on the electric motors currently used to saw, drill, print, press, mold, and spin materials—applications that collectively account for about another 10 percent—and once again, the new technologies can do these jobs far more efficiently than the old. But if experience is any guide, overall energy consumption will not drop as a result—new applications will multiply faster than new efficiencies are incorporated into the old ones.

At the end of the day, a burgeoning catalogue of new technologies does not sound like it can provide any kind of useful guide for public policy. But that's the point—no one in 1980 could have foreseen the next two decades of light-bulb evolution, and it is no easier to look more than a few years ahead today.

Only two reasonably clear lessons emerge from this history.

First, efficiency seems to come, regardless—often far more efficiency than the most well-meaning regulators and policy pundits can foresee. While the negawatt faithful were thrusting fluorescents on users who could hardly be persuaded to take them for free, greedy capitalists were developing far better semiconductor-based bulbs and amplifiers. It rarely makes any sense for regulators to try to promote "more efficient" technologies, because given the fecundity of technology, there's no reason to suppose that regulators will reliably choose the right technologies to promote, or the right time to promote them.

Second, when radically more efficient technologies do emerge, they are quickly embraced by paying customers without any need for government mandates—embraced not just to displace old ways of doing things, but to do all sorts of new things that previously hadn't been done at all. Which means, at the very least, that rising efficiency certainly does not guarantee falling energy consumption. Through all of technological history on record so far, it has had just the opposite effect.

THE PARADOX OF EFFICIENCY

Nearly everything said about a new car means lower fuel efficiency—for example, larger body, longer wheel-base, greater weight, softer tyres, more horse-power, more rapid acceleration, higher speed, automatic drive, improved flexibility of control. And now powered steering is almost upon us! No wonder car miles per gallon of gasoline have shown no improvement over the past thirty years. Passenger-miles per gallon have probably shown a definite decline, because an even larger proportion of motorists like to ride alone in about two tons of assembled steel driven by a silent power plant capable of more than a hundred horse-power. This decline is in spite of engineering changes that have made considerable improvements in ton-miles per gallon.
    —Eugene Ayres and Charles Scarlott, *Energy Sources:*
                    The Wealth of the World (1952)*

A certain amount of waste may be thermodynamically inevitable, but excess losses can nevertheless be curtailed. And it seems obvious that

*McGraw-Hill, 1952. This is probably the first modern book devoted entirely to the subject of energy; it is a work of outstanding depth, rarely matched, still less surpassed, in many decades of subsequent analysis. The authors report that they began writing when—to most everyone's surprise apparently—the subject of "energy sources kept coming up as a topic for discussion" in a 1948 meeting of the American Petroleum Institute.

rising efficiency in cars, furnaces, and lawn mowers should, in the aggregate, significantly curb demand for energy. Sad to say, however, the chronicles of light discussed in the preceding chapter reflect a much broader truth: efficiency doesn't lower demand, it raises it. That the pursuit of efficiency has been the one completely consistent and bipartisan cornerstone of national energy policy since the 1970s only strengthens the conclusion. Efficiency has come, and demand has risen apace.

THE EFFICIENT FUTURE

Still more efficiency lies ahead, in every major sector of our energy economy. Much of it will come from the rise of the hybrid-electric power train that we discussed in chapter 5; the same technologies will, in tandem, greatly improve the efficiency of the machines that power factories.

Much like the car, those machines remain far bulkier and heavier than they ought to be. In a robotic welder, the wires that actually apply the intense heat to the metal are tiny; most of the bulk lies in the tangle of mechanical/hydraulic control systems that move the electric arc to the right place. Devices that fill boxes, tighten screws, move liquids, snap together parts, and automate assembly lines in countless other ways are still, almost invariably, far bulkier and heavier than the job ought to require. The new power semiconductors displace pounds of click-click mechanical and hydraulic hardware with ounces, or less, of electrical wiring and semiconductor chips. They are being embraced in factories, as in Detroit's engine-design shops, not because federal energy policy so requires, but because they are lighter, perform better, and are also (though last and least) much more efficient.

They pollute less, too. The steam regulators and ignition systems developed long ago by James Watt, Etienne Lenoir, and Nikolaus Otto, and only improved at the margin in the many decades since, are at last giving way to completely different, dexterous designs—highly intelligent robots, really—embedded deep inside our engines. Commercial camshaft-free electric-valve engines are just beginning to appear on the highway; they lower emissions by 50 percent or more even as they raise efficiency by 10 to 20 percent. Unlike scrubbers and catalytic converters, which pursue pollution after it is formed and increase fuel consumption in the process, these control systems operate mainly at the front end of the energy conversion process, and lower emissions even as they raise efficiency.

The tuning of a complete vehicle can be pushed much further by electrical drive trains. Torque, traction, braking, skid control, and fuel economy all depend on the complex interaction of engine, battery, suspension, steering, and brakes; the magic lies in the intelligent coordination of all the parts. Emissions depend on exactly the same interactions; the intelligence that is added to improve performance along every other dimension can readily improve performance along green metrics too. Large (10,000-unit) packs of super-capacitors, for example, are now being incorporated in the drive trains of hybrid buses and trucks. These ultralight power caches pack enough punch to accelerate and drive a city bus several blocks before the small, highly efficient, ultra-clean diesel kicks in, both to recharge the capacitors and to add torque to the wheels. Emissions drop 90 percent and fuel efficiency rises 60 percent because engines run so much better at steady speed and load.

Similar improvements are possible wherever combustion and thermal processes are used to purify, synthesize, and treat materials. The billions of electric valves that currently control gas and liquid flows in factories and homes are being linked to sensors that monitor temperatures, pressures, flow rates, and other variables. The chemical factory gets networked end to end, and dexterous robots built into the hardware take control of the boiler, the furnace, the mixing vat, and the reaction vessel. By far the most effective way to raise efficiency and to control pollution is to improve control of the flow of energy through the systems that consume it.

Finally, as discussed in the previous chapter, the efficiency of electron-to-photon conversion technologies is improving faster still, as white-hot filaments give way to semiconductor junctions. LEDs already offer far better performance than either incandescent or fluorescent bulbs. Here too, the efficiency gains come with better performance, not worse. The laser diode advances solid-state light to the point where it can displace the conventional thermal and material processing technologies. A fiber-optic system requires far less power to transmit bits than an electric wire. Microwaves can heat just the water in the soup, not all the air and stove-top around it, and lasers deliver similar improvements, only more so.

THE EFFICIENT PAST

With so much new efficiency so close at hand, why then do we doubt that efficiency will curb energy consumption? The short answer is that the technologies are new but the trends are not. As Hans Thirring recounted in his remarkable 1958 book *Energy for Man*, efficiencies have been rising for as long as there have been thermal engines.* Two centuries ago, no engine could surpass 10 percent efficiency. By raising boiler temperatures and pressures, engineers pushed performance to about 20 percent efficiency by the turn of the twentieth century. By mid-century, they were up to about 40 percent. Today, the best thermal plants routinely hit 50 percent efficiency.

Efficiency gains this large ought to have had a dramatic impact on supply and demand—and they did. The price of transportation and electricity fell steadily. And the total amount of fuel consumed in those sectors rose apace. Efficiency may curtail demand in the short term, for the specific task at hand. But its long-term impact is just the opposite. When steam-powered plants, jet turbines, car engines, light bulbs, electric motors, air conditioners, and computers were much less efficient than today, they also consumed much less energy. The more efficient they grew, the more of them we built, and the more we used them—and the more energy they consumed overall. Per unit of energy used, the United States produces more than twice as much GDP today as it did in 1950—and total energy consumption in the United States has also risen three-fold.

---

*Hans Thirring, *Energy For Man: From Windmills to Nuclear Power* (Indiana University Press, 1958), p. 157. Thirring may have been the first energy analyst to take specific note of energy consumed in telecommunications and broadcasting. That sector, he estimated, consumed about 0.6 TWh of electricity in 1958.

FIGURE 7.1   Energy Cost of Transportation versus Total U.S. Consumption*

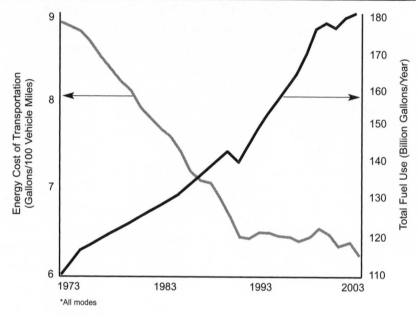

*All modes

Source: Department of Transportation, *National Transportation Statistics 2003*; EIA, *Annual Energy Review 2003*.

Efficiency improvements have not lowered the amount of fuel consumed in transportation. The amount of fuel needed to move a vehicle 100 miles has fallen steadily, but total fuel consumption in the transportation sector has gone up.

FAST

First of all, efficiency fails to curb demand because it lets more people do more, and do it faster—and more/more/faster invariably swamps all the efficiency gains.

Eugene Ayres and Charles Scarlott summarized the simpler version of this story half a century ago, in language that every new generation of energy pundits seems fated to repeat. "Higher automotive-engine efficiencies are announced from time to time as resulting from improved engines or superior fuels. But motorists have not realized any increase in mileage, since the potential efficiency increase has been offset by running more powerful engines under lighter partial loads and in hauling more tons of automobile at higher average speed. The motor-car operator is

FIGURE 7.2   Energy Cost of the U.S. Economy versus Total Consumption

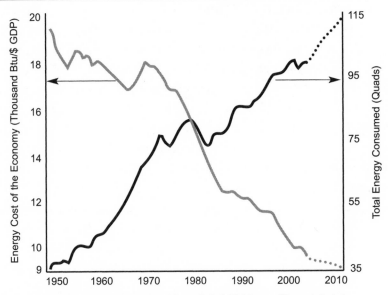

Source: EIA, *Annual Energy Review 2003* and *Annual Energy Outlook 2004*; Bureau of Economic Analysis.

The U.S. economy as a whole is twice as energy-efficient today as it was in 1950—the amount of fuel needed to produce $1 of GDP has been cut in half. But total energy consumption has almost tripled. As they grew more efficient, we built more steam power plants, jet turbines, and car engines, light bulbs, electric motors, air conditioners, and computers, and used them more heavily—and total energy consumption went up.

largely to blame. He insists on excessive weight, power, and luxury. He accelerates too rapidly. His speed is excessive. He knows, in a vague way, that he is paying for all this, but he feels that he is getting his money's worth of what may be called the 'amenities' of motoring."[1] The problem, however, is even more fundamental than that. Ridiculous though the proposition may sound, the technologies that boost efficiency are those that burn fuel faster, in less space. And these technologies invariably boost overall demand, as well.

It is of course true that if we simply replace an existing light bulb or engine with a more efficient one, and change nothing else, we cut energy consumption. But very little of this kind of active retrofitting ever happens —homeowners, at least, don't generally replace light bulbs until they burn out or junk cars until they're ready for the scrap heap. In the short term, efficiency gains generally impel us to add to rather than replace.

That's the heart of the problem. To reduce energy consumption, a more efficient technology has to have a greater impact in the replacement market it creates than in new markets it infiltrates New, more efficient engines must replace old ones faster than we find new uses for the new-and-improved engines. LEDs have to replace old light bulbs, for example, faster than they get deployed in jumbotrons and countless other places that the very compact, cool, new light can go—all the new applications that old bulbs couldn't serve at all. But this just doesn't happen. The new uses invariably multiply faster than the old ones get retrofitted.

As a general principle, more efficient devices are more efficient because they run *faster*. But faster devices get used more, to deliver more miles, generate more electricity, weave more fabric, or reap more wheat.

Faster thermal systems are more efficient because the main inefficiencies come from heat losses to the environment, and the faster the engine runs, the less time there is for thermal leakage. Gas turbines are highly efficient, but only when their blade tips move at near sonic speeds. When Robert Goddard began experimenting with rocket engines in 1915, he adopted the nozzle that Gustaf de Laval had designed to drive the first steam turbine. Obtaining jet velocities above 7,000 feet per second, he pushed the efficiency of his rockets up to 63 percent—making them the most efficient heat engines ever built, as Goddard himself proudly noted. This may seem ludicrous to anyone not steeped in thermodynamics, but it is self-evident to anyone who is: the theoretically attainable efficiency of a heat engine increases with its operating temperature; the faster you burn fuel the hotter it gets; and no heat engine runs hotter than a rocket's.

If all our intuition still insists that rockets and other high-speed thermal engines simply *must* be less efficient, it is because we intuitively mix up two quite separate concepts, efficiency and burn rate. Efficient engines generally run fast, get a lot done fast, and burn fuel fast. Inefficient engines are inefficient *because* they run slowly, and they run slowly because nobody has yet figured out a faster—and thus more efficient—design.

FIGURE 7.3   Combustion Engine Efficiency and Speed

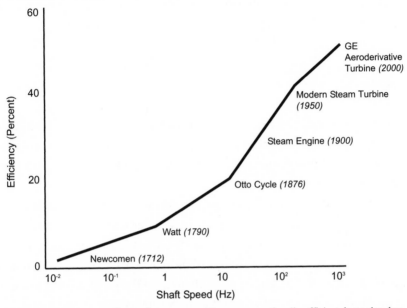

More efficient devices are usually run faster—gas turbines are exceptionally efficient, but only when their blade tips move at near sonic speeds. But faster devices get used more, to deliver more miles, generate more electricity, weave more fabric, or reap more wheat.

Why then are we repeatedly told that driving slowly "saves fuel"? It does, but only because it wastes time. Lowering the speed does indeed lower aerodynamic drag on the vehicle, but people drive faster for a reason —to get somewhere sooner. "Efficiency" is supposed to save fuel by doing the same job better; it is always possible to save fuel by doing less of a job, worse. When it comes to travel, time spent on the road matters as much as distance covered, and it's silly to talk of "saving fuel" without acknowledging that something important is being wasted in its place.

Because it's so easy to fiddle with the accounts, all serious measures of efficiency focus on precise and solid engineering metrics of output, like rocket thrust or shaft power. Rocket engines are very fast, and thus very efficient at what they're designed to do. But rockets don't "save fuel," they consume it. This is just another way of saying that efficiency doesn't lead directly to "saving energy," and it doesn't lead there indirectly, either. The only clear conclusion is that more efficient almost invariably means faster, and faster almost invariably means a higher burn rate, and more miles traveled, and more energy consumption overall.

LIGHT

For things that move, efficiency also rises when engines get smaller and lighter, because less of the engine's output is then used to haul around the engine itself. But advances that make things smaller and lighter invariably open up lots of new uses, too.

Bell Labs developed the transistor specifically to replace the triode vacuum tube, which was too big, hot, and inefficient to use in setting up telephone connections. The transistors on today's integrated circuit are a hundred million times smaller and more efficient, but we manufacture quadrillions of them, and most of them aren't used in telephone circuits at all. In the aggregate, they use far more energy.

Similarly, there's no doubt that cars will get much more efficient as Detroit adopts the electric drive train. But on-board the SUV, the electric drive train will also power a host of new capabilities that we don't yet enjoy at all. The existing suspension is just a passive absorber of shocks; combine electrically activated suspension with suitable look-ahead sensors, and the suspension can see the pothole before the tire hits it, and move the tire down and back up with just the right timing to eliminate the jolt entirely.

And once Detroit embraces the new technology, the electric drive train will get very much cheaper. As a direct result, scooters, lifters, and robots will multiply in new niches that aren't currently powered at all. Dexterous robots will take over more and more of the lifting and moving of our daily lives, replacing human power with machine in the factory, office, and home, much as horsepower on the highway was replaced by the automobile.

The Segway scooter noted in chapter 5, for example, has been touted as a replacement for cars. But the new scooter is much more likely to get loaded into the back of the old SUV to save the effort of trudging around the mall, campus, or resort. Fly by wire controls cut weight out of commercial jets; they also make possible pilotless Predators that can fly reconnaissance for the military around the clock and around the globe. In chapter 8 we describe the da Vinci surgical operating system, an electrically powered system that stands between surgeon and patient. The device doesn't just substitute for the surgeon's own muscle power, however, it makes possible a limitless range of new interventions that would never be feasible or affordable without the assistance of the new machines. The machines do gall bladders; they also do liposuction.

Advances in solid-state light will have much the same impact, on both efficiency and demand. LEDs are already displacing incandescent bulbs in cars, inside the passenger cabin, and in the brake lights, to begin with—and certainly saving energy as they do. But as discussed in chapter 6, the new light bulbs in the car won't emit just light—in high-end vehicles they already project millimeter waves for active cruise control, or sense infrared light for enhanced nighttime vision. More bands of light and sight are coming, and they will inevitably end up in less expensive cars too. And in our homes and in public places as well, to make life safer, more comfortable, more convenient—just as light bulbs did when they were introduced over a century ago. The new solid-state light bulb doesn't just supply a more frugal desk lamp, it does laser surgery and makes it affordable for the masses.

The new technologies are efficient because they are quick, compact, light, and responsive. Those same attributes open up new vistas of responsive, intelligent, delicate, high-precision motion that weren't possible before. To make the same point retrospectively, all the important opportunities for new efficiency lie in reducing the losses in energy-consuming technologies that have already been embraced by the mass market. No one could improve gas mileage until there were gasoline engines, and policy makers didn't get interested until Detroit was building tens of millions of them. Even then, efficiency is only improved at the margin, when old cars, light bulbs, and motors wear out and get replaced by new ones. At the top of the pyramid, by contrast, everything is new, and every day we raise the apex a little higher, as we discover new ways to use new forms of power. New miles, new types of demand multiply and take hold faster than the new efficiencies do.

Small size, it turns out, has the same overall impact on fuel consumption as high speed. If you can make an engine smaller without losing more heat through its surface—that is, if you increase power density by burning more fuel, faster, in less space—you improve efficiency and simultaneously accommodate new demand. On the single metric of overall fuel consumption, the efficiency more than defeats itself.

FASTER AND LIGHTER

Perhaps, however, the information economy changes things so fundamentally that—for the first time in history—rising efficiency will outpace rising demand. One bold theory holds that it will. In the new Harry Potter economy we won't have to drive to the mall to pick up the latest kids' bestseller; we'll order it from Amazon. And Amazon's business model requires less energy than Barnes and Noble's. An economy that moves bits rather than atoms just won't need as much energy.

The argument runs as follows. GDP historically grew in lockstep with energy consumption, but it no longer does. "Energy intensity"—the energy consumed per dollar of economic output—is declining, because the new wired economy is so much more efficient than the old. Online stores substitute "clicks for bricks." Delivery trucks are more efficient than car trips to malls. Wired supply chains reduce inventories, cut overproduction, reduce unnecessary capital purchases, eliminate paper transactions, reduce mistaken orders, and thus save energy all around. The Internet will eventually cut travel, too. GDP growth is thus being decoupled from energy consumption entirely. The really big gains in efficiency that the new technologies make possible will first stabilize, and then roll back, our total consumption of energy.*

*Ongoing research and a broad body of publications along these lines are maintained through federal funding at the Lawrence Berkeley Labs, eetd.lbl.gov.

The most basic problem with this argument is that energy consumption per unit of GDP has indeed been falling pretty steadily—not since the advent of the Internet but for the last twenty thousand years. Any different conclusion is based on trend lines that ignore the carbohydrate tiers of the energy economy, which continued to play a large role even in industrialized countries until well into the twentieth century. The GDP of our hunter-gatherer ancestors was nothing but energy, in the form of food calories. Wealthier economies add less energy-intensive goods to the mix, so energy per GDP falls: it doesn't take much oil to write software or to fuel a symphony orchestra. But this "dematerialization" of the economy is entirely relative. Our consumption of knowledge-intensive goods grows faster than our consumption of energy-intensive goods—but both continue to grow. Adding a lot of energy-lite consumption is like adding artificial sweetener to a diet—it doesn't, in itself, lower total calories consumed.

As we discussed in chapter 2, the bit-moving technologies require surprising amounts of energy too, in both their manufacture and their operation. The question, then, is whether the energy the new information technologies consume is more than offset by the energy they save.

The Web's main impact is to move everything faster through the entire length of the supply chain, from the forest, farm, mine, or well, down to the consumer. This does make supply chains much more efficient, but here again, the efficiency is a direct consequence of higher speed—by speeding up the flow of information about supply and demand, we speed up the flow of goods. Suppliers can now provide instant gratification, and buyers quickly come to expect it. In the summer of 2003, Amazon had to hire a fleet of Federal Express jets to deliver a new Harry Potter release to all its customers overnight. It is always possible that—through all sorts of secondary effects—moving goods faster through the pipeline will save energy overall. But if so, then higher speed limits on the highway ought to save fuel as well—and there is, unfortunately, no reason to suppose that they do.

How efficiency improvements in one sector of the economy—telecommunications, in this instance—are likely to affect energy consumption in others is not a question one can ultimately answer by theorizing.

FIGURE 7.4   Cost of Transmitting Information versus Total U.S. Energy Demand

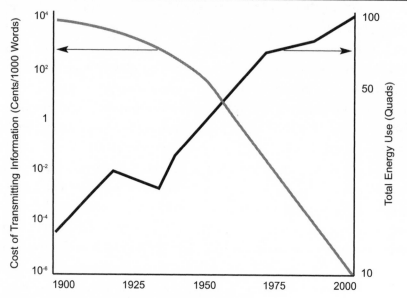

Source: EIA, *Annual Energy Review 2003*; U.S. Census Bureau, *Historical Statistics of the United States Colonial Times to 1970*; Ithiel de Sola Pool, *Technologies of Freedom* (Bellknap Press, 1983).

Historical trends do not support the suggestion that improvements in information technology will reduce demand for energy. The cost of conveying information has been dropping rapidly since the days of Bell and Marconi. But energy consumption has risen relentlessly.

Historical experience is the best thing we have to go on, however imperfect a guide it may be. The cost of conveying information electronically—and thus, fast—has been dropping rapidly since the days of Bell and Marconi. And with faster information have come faster cars, jets, and power plants, and more energy consumption all around. When we learn more, faster, about the world around us, we engage it more, not less. There is then, perhaps, reason to suppose that higher speeds on the information highway will raise—not lower—energy consumption. As Vince Cerf, one of the fathers of the Web, has observed, "the Internet has the funny effect of increasing the amount of travel—people using it discover places to go and people they want to meet."[2]

INEFFICIENT BY DESIGN

Information and telecom technologies aside, nowhere have we improved efficiency faster than in the technologies we use to extract and capture energy. Agrarian societies use almost all the energy they produce in clearing and cultivating the land where they produce it. The energy required to extract energy itself has fallen dramatically as we have found ways to displace 6 tons of wood with 3 tons of coal and then with a few grams of enriched uranium.

And here, as elsewhere but even more so, rising efficiency has led to more energy consumption, not less. From wood to coal to oil to uranium, new fuels packing more energy into less space could be distributed more quickly and conveniently, and burned faster, so they were. Altogether new forms of demand materialized around the new fuels. Coal extracted initially to replace wood for heating found much larger markets in steam engines, and then electric power plants. Oil initially used to fuel cars found other large markets in ships, jets, and chemical industries. Electricity initially supplied to power Edison's new bulb was soon tapped to power electric motors, then compressors in refrigerators, then air conditioners, and then microprocessors.

The efficiency of fuel extraction plummets, of course, when we push things in the opposite direction, from nuclear and fossil fuels back to carbohydrates and the sun. The commonly supposed efficiency of "alternative" technologies and fuels is not tied to any standard engineering metric. When a Brazilian farmer burns down 2 acres of rainforest for pasture, where should his energy-efficiency accounting even begin? With the solar energy falling on his field, or what gets captured by the grass, or digested by the horse? Some 99.999 percent (or so) of the incident solar doesn't make it through to the horse-drawn plow, and some 98 percent of the tiny remainder that does never makes it into calories that the farmer himself ultimately digests.

What about the fuel cell—so elegant in its conception, so clean in its actual operation, and in such favor among those who despise internal combustion? Is *it* efficient? Perhaps. If used to generate electricity in homes, the cell's waste heat can be used for space heating and hot water. We use some 40 percent of all our fuel to generate electricity, and about 30 percent for heat, so there is a possibility of some real saving in bringing the two together.*

As anticipated in chapter 5, however, the fuel cell's efficiency ultimately depends on how we get the hydrogen to fuel it. Hydrogen is abundant in water, but water is what you end up with *after* you burn it. The only readily available supplies of hydrogen in relatively "unburned" form are in hydrocarbons, which are, of course, part hydrogen, part carbon. Both components burn well—charcoal barbecues and blast furnaces burn pure carbon, the Space Shuttle's main engine burns hydrogen. Over half of the heat from a methane-gas flame comes from the hydrogen, but only about 20 percent of the heat in coal. If we make methane gas our prime source of hydrogen, we will, in effect, only burn about half the combustible fuel and discard the rest. This isn't "efficient" at all, except perhaps politically, in that it forces a shift in fuels without confronting the coal industry directly. The great virtue of the fuel cell's PEM, it turns out, isn't efficiency, it's delicacy—the PEM is the canary in the mine that can't stand even a whiff of carbon, and therefore despises coal.[†]

The efficiency numbers are equally bad, or worse, if we use electricity at the outset to extract hydrogen from water. As already noted, the complete cycle of water to hydrogen to water consumes four times as much electricity as it produces. The case for harnessing any part of our energy economy to this cycle centers on the assumption of "free" solar

---

*A lot less than these two numbers initially suggest, however. A home generating all its own electricity would produce far more waste heat than the home can use for heat and hot water. Many industrial factories, by contrast, require much more heat, and at much higher temperature, than can be obtained by co-generating electricity on the premises.

[†]As things currently work, the carbon stripped out of the gas by a "reformer" placed upstream of the PEM cell is simply dumped back into the air as carbon dioxide, anyway. There are other fuel cell designs that run hotter, are far more robust, and can use hydrocarbons directly, but they attract far less interest.

(or nuclear) electricity, with the hydrogen simply used for highly inefficient storage.

EFFICIENT WASTE

It is only when we begin to focus on efficiency in the extraction of energy that the paradox of efficiency comes to seem less paradoxical. Of course energy consumption rises as we grow more adept—more efficient—at extracting oil, coal, and uranium from the ground—or, for that matter, at plucking electricity out of the sun and wind. The better our energy-extracting technology, the cheaper the energy, and when goods get cheaper, we consume more of them. There's nothing paradoxical at all about *that* proposition.

The only thing one must then add is that almost all of the efficiency debate is about extracting and plucking. As we discussed in chapter 3, well over 90 percent of all the energy we use is used in mining, drilling for, refining, processing, and converting energy itself. This doesn't happen just at the well-head or the strip mine; it happens in the furnaces that transform fuel into heat, and in the engines, cars, boilers, and turbines that transform heat into motion, and in the generators that transform motion into electricity, and hence through the motors, bulbs, and refrigerators, to the last, comparatively minuscule trickle of highly refined energy that we actually end up putting to real use.

From the end user's perspective, new efficiency in any of these intermediate tiers has exactly the same economic impact as new efficiency in the technologies that extract oil from 6,000 feet under the bottom of the ocean. Small wonder, then, that efficiency increases consumption. It makes what we ultimately consume cheaper, and lower price almost always increases consumption. To curb energy consumption, you have to lower efficiency, not raise it. But nobody, it seems, is in favor of that.

POWER, PRODUCTIVITY, JOBS, AND GDP

Those mechanical wonders which in one century enriched only the conjurer who used them, contributed in another to augment the wealth of the nation; and those automatic toys which once amused the vulgar are now employed in extending the power and promoting the civilization of our species.
—Sir David Brewster, Letters on Natural Magic (1835)[1]

If the technologies of power were now at a standstill—if our energy future were only a shrinking version of its past, with energy prices rising, the capabilities of engines, motors, and machines frozen in place, and energy consumption per American worker in decline—the future of employment in the United States would be bleak.

Japan and Germany would not be the concern: their populations are shrinking, neither will take effective steps to expand energy consumption, and as a direct result, their economies will stagnate. Collectively, European countries currently use only half as much energy per capita as America does. Some part of that difference is attributable to higher population density (four times higher than in the United States, higher even than China's) and comparatively mild climate; the rest reflects limited resources, bad energy policy, and geopolitical weakness. Rather than confront and reverse their decline, they portray the energy they don't have and the power they don't use as matters of deliberate choice, made to save the environment. Overwhelmed by lethargy, they now pronounce it virtuous.

But the emerging economic powers of Asia, and perhaps India too, are quite different stories. Their workers are now rapidly gaining access to the powered machines that steadily raised the productivity of the American factory worker throughout the twentieth century. The labor of their factory workers is now amplified by electricity generated in power plants designed and often built by U.S. vendors, but at much lower cost. Their power is a lot cheaper than ours, because they're getting it the old-fashioned forget-the-environment way, just as we used to ourselves. There is little fuss or bother about protecting the river or scrubbing the smoke. China's answer to the 2-gigawatt Hoover Dam on the Colorado River is an 18-gigawatt dam on the Yangtze River.

Combine much cheaper supplies of energy with ready access to heavy industrial machines and it is hard to see how these countries can fail to close the productivity gap that has historically permitted American factory workers to remain competitive at much higher wages. Unless perhaps another industrial revolution is at hand, to vault the American workplace far out ahead of the competition once again.

It is. The first revolution—Watt's—transformed American industry. The second—Otto's—transformed American transportation. The next great revolution in powered machines—the one we owe to Shockley, Becke, and Wheatley, the one defined by the rise of digital power, dexterous robots, and networked factories—is already well under way, and is transforming both, with America leading the way. In the industrial sector, the new machines don't move bits, they move stuff—out of the American mine and the farm, through the American factory, along the American assembly line, down the American highway, over the water, and through the air. The machines are nimble, thoughtful, responsive, and intuitively dexterous. They mimic and then improve upon the remarkable array of muscles, sensors, and intelligence that allow the human body to waltz, play the violin, or wield a hammer. They amplify and enhance not only raw power but thoughtful, adaptive power, in ways that are now boosting the productivity of our workers far beyond anything ever imagined before.

These machines, in short, are delivering the next great boom in productivity, employment, and national wealth. And happily for America, we are well out ahead in designing, building, and deploying the machines themselves and the power semiconductors and software from which they are built. So long as U.S. manufacturers continue to widen that lead, American workers will readily hold their own against any labor force on the planet. Indeed, the competitive advantage in manufacturing is now swinging decisively back toward the United States. The rise of American industry was what made the twentieth century the American century. If we remain willing to seek energy, use power, and embrace new power-controlling technologies, the twenty-first century will be the American century too.

POWER AND PRODUCTIVITY

Power is one of three fundamental inputs that determine the productivity of labor in every sector of the economy, the other two being material and information. Capital investment—often emphasized in the economic analysis of labor productivity—typically looks to intelligent configurations of concrete, steel, and silicon. But power is what brings those structures to life. In agrarian societies, the capital investment goes mainly into the clearing of land for agriculture; the power trickles in from the sun. In modern industrialized societies, cataracts of power are extracted from coal, oil, uranium, and natural gas. As Lewis E. Lehrman has noted,[2] the historical correlation between rising employment and rising consumption of these now conventional fuels is simply too strong to deny. It's energy that gets the increasingly mobile worker to the increasingly distant workplace, and energy that processes material and powers the increasingly advanced machines that shape and assemble it.

It has always been thus. For half a millennium, as Jean Gimpel describes in *The Medieval Machine*, water mills played a key role in the economic development of England. The Doomesday tax collectors dispatched

FIGURE 8.1   Productivity, Energy, and the Economy in the United States

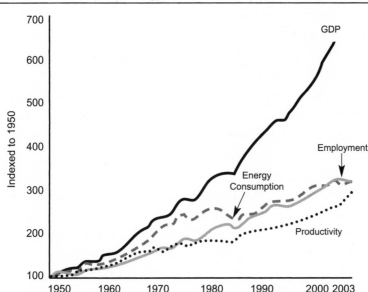

Source: EIA, *Annual Energy Review 2003*; Department of Commerce; Department of Labor, Bureau of Labor Statistics.

Power is one of three fundamental inputs that determine the productivity of labor in every sector of the economy; the other two are material and information. Capital constructs intelligent configurations of concrete, steel, and silicon; power makes these structures run. Power moves the worker to the increasingly distant workplace, drives the intelligent machines that surround him there, and processes the materials and information that he works with.

by William the Conqueror in 1086 recorded 5,624 water mills, or about one for every fifty households. On the far side of the channel, windmills became so important that Pope Celestine III (1191–98) taxed them.[3]

While water mills still ground the corn, human hands supplied most of the dexterity and power in the factory. The word *manufacturing* comes from *manufactory*, meaning "made by hand." Then, quite abruptly, everything changed. In 1804, scarcely three decades after Watt invented his engine, Blake's preface to *Milton* evoked "dark satanic mills" looming up against the English sky. The future of industry was now defined not by human, horse, and the water-powered grinders of corn, but by steam-powered weavers of cloth. Originally invented to mine its own fuel, Watt's steam engine and its fuel together were now being enlisted to drive the looms in textile mills. Burning coal now somehow managed to weave delicate fabric, loop by loop, stitch by stitch, with the former weaver now stoking the furnace and tending to the machine.

Belching fire and smoke, the new infernal machines promised prosperity to a few but seemed to threaten the livelihoods of many more. Fearful of what lay ahead, the Luddites—whose muscle-power and dexterity were at peril—broke into factories to smash the automated looms. As Kirkpatrick Sale describes in *Rebels against the Future*, they had neither an articulated strategy nor the means to stop the inexorable march of the machine. After two years of agitation they disappeared.

Employment somehow survived the rise of steam power, however, and would go on to expand through each subsequent revolution in the use of fossil fuel to displace the power and dexterity of hand and muscle. By 1920, powered looms had made the British textile worker so productive, and the textiles he spun so cheap, that his cloth could be sold at a profit even to the impoverished farmers of the Kathiawar peninsula. In campaigning for his country's independence, Mahatma Gandhi exhorted Indians to spin their own cloth instead, on their traditional, hand-powered cottage looms.

By 1920, an altogether new water mill, this one powered by Niagara Falls, was coming to the fore—to generate electricity. Standing alone, the steam engine had provided only mechanical power—the power of the nineteenth-century factory. Quite suddenly, right around the turn of the twentieth century, factories were evicting steam engines from their premises; by 1910, as David E. Nye recounts in *Electrifying America,* the transition to the electric factory was well under way. Factories didn't want *energy*, it turned out, they wanted *energetic order*—and electricity in wire supplied far more of it than torque from a steam engine's shaft.

For a while, generators in the Niagara power houses extracted the new power from falling water, but before long steam engines were back in the picture. Now, however, they weren't in the factory; instead, they turned generators situated miles away. Edison's biographer Matthew Josephson estimates that this transition to better-ordered power driving smaller, faster, electric machines quickly boosted assembly-line efficiency by 50 percent.[4] As traced out in a landmark 1986 study by National Academy of Sciences, interposing an electric generator between the combustion engine and the workplace changed the nature of work, and amplified the productivity of the worker, as much or more than the original combustion engine itself.[5]

Today, another fundamentally new layer of power-processing technology is being interposed between the generator and the payload. The technology is profoundly new and different; there is thus every reason to suppose that it will transform the workplace once again, and again to the long-term advantage of the worker. Watt's reciprocating-piston steam engine made possible the industries of the nineteenth century, but it could never have made possible those of the twentieth. The second industrial revolution—the twentieth century's—was propelled by much smaller, faster machines—Otto's internal combustion engine, de Laval's steam turbine, and the Tesla–Westinghouse electrical generator. Those machines are now giving way to the technologies of digital power—machines built around power semiconductors and electric drive trains.

In retrospect, at least, it's always obvious: supply the worker with more, better-ordered power, and he produces more; if the new machines are different enough, both the factory's and the worker's productivity improve beyond recognition. Gandhi was right about culture and politics, but he was dismally wrong about basic economics—his hand-powered looms stood as much chance against powered looms as the Zulu's assegai throwing spears did, when bravely hurled in the direction of ten-barrel, 320-round-per-minute Gatling guns.

DEXTEROUS ROBOTS

Until quite recently, machines added dumb power or simple-minded precision to our lives. A steam shovel could certainly lift more dirt than a man with a spade. Lithographic machinery could etch much finer lines on a chip than human fingers ever could. But machines couldn't ride a bike or walk up stairs, and they didn't make good brain surgeons either.

Robots were all the rage in 1980. Japan had mastered them, we hadn't, and they—the robots and the Japanese both—were going to take over all of manufacturing, and thus all the heavy lifting in the economy.

But for all they did, the robots of that era turned out to be cost-effective for only the very largest manufacturers—some two-thirds of the orders in the 1980s came from the automotive industry. These robots were suitable only for highly repetitive tasks, because setting one up to perform a particular task was a long, complicated process. Most of the rest of "factory automation" consisted of small versions of the same—simple switches and turnstiles, for example, that could sound an alarm when a box on a conveyor belt fell on its side. As a result, U.S. orders for robots peaked at just over six thousand in 1985 and dropped steadily for most of the following decade.[6]

The machines now emerging are immeasurably better. Consider, as a first example, the da Vinci surgical system. Like an astronaut operating the Space Shuttle's robotic arm, the da Vinci surgeon sits at a console and monitor, with his hands on seemingly familiar instruments that move and push back just as they would if he were reaching directly into a thorax or a skull. And that is where his delicately moving fingers are indeed reaching—but through a keyhole incision in a patient several feet away.

To take surgery from open-chest to keyhole, the mouse and computer had to evolve to the point where they became a complete surgical console—a flight-simulator for surgery, with the surgeon able to handle instruments and view a screen in ways that look and feel just like he's operating directly on the real thing. At the other end of the surgical-robotic arm, sensors had to evolve to the point where they could not only video the real thing, but reliably transmit all the tactile information that the remote surgeon needs at his fingertips. Tiny motors—backed by equally compact power supplies—had to evolve too, to the point where they could do the cutting, aspirating, and stapling through the keyhole. But all this did happen, and is happening—in the operating theater and in countless other workplaces. Coal, burning in a furnace a hundred miles away, now moves the blade that repairs the human heart.

That the robotic surgical arm requires only a tiny incision is one huge advantage. A second is that the machine-controlled surgical instruments are now even more delicate and finely coordinated than the human hands behind them. The computers in the middle can filter out every vestige of tremble in the surgeon's fingers. With the aid of microscopic, hyper-sensitive pressure sensors, they can also learn to feel the boundaries of blood vessels and tumors even better than the surgeon's hands alone ever could. And with the help of high-resolution MRI maps or even real-time imaging, they will soon begin acquiring self-navigation skills.

Why could no one build such a system even a decade ago? What do we have today that was missing back then? Three key layers of technology: motors and electric power supplies compact and precise enough to mimic the small muscles of a hand; sufficiently fast and accurate sensors to provide real-time feedback of what's happening at the payload; and sufficiently fast computers to make sense of it all, and to constantly re-calculate how much power to dispatch to the motor to get the device to climb the stairs or replace a heart valve.

Those technologies, however—two of the three being core technolo-gies of power—are now at hand. And because they transform the human-machine interface, they represent, ironically, a far bigger advance than fully autonomous robots. Like the Segway scooter, da Vinci is as quick and nimble as the human behind it. Such machines feel what our own brains don't even consciously grapple with—mental instructions con-veyed directly, and very fast, to our finest muscles—and respond as fast and as delicately as our own bodies do. Machines that can do this define a new world far beyond "robots." They reflect the convergence of digi-tal logic and digital power.

DIGITAL FACTORIES

The place to see the digital robot already ubiquitously at work—and hence to see the future of the factory more generally—is in a chip fab. The linen, hemp, and cotton spun by the machines the Luddites feared have given way to silicon, gallium, and germanium. Worsteds, woolens, damasks, and brocades have become microprocessors, RAM chips, DSPs, MOSFETs, and IGBTs. We still weave cloth too, but the weaving of atomic-scale semiconductor fabric is a now a vastly larger and more important industry.

The devices that arrange matter at these scales are so precise they are generally called "instruments" (or "tools") rather than "motors" or "machines." Wafers are mounted on digitally powered "planetary motion" platters, which rotate on command. Molecular, ion, and chemical-vapor beams are directed at crystalline surfaces, where they react and grow, in thicknesses sometimes measured in the tens or hundreds of atoms. No functioning PC emerges without the most accurate control of the chemical purity, crystalline structure, position, speed, and acceleration of silicon itself and of the systems that dope, etch, slice, and package it. The finely etched electrical connections and transistor features must be aligned to submicron levels of precision. Fantastically clean and precise motors and lasers move crystals, bond wires, probe wafers, trim edges, drill and inspect circuit boards, and pick and place components onto them. Few Luddites are at hand to attack the machines that weave semiconductors; people are too dirty and clumsy to be allowed anywhere close to the crystals and the atomic-scale junctions that are being erected upon them. Humans could never perform such labor by hand in any event.

Yesterday's machines couldn't either. Slicing up a silicon wafer, to pick one of the simplest tasks, requires a motor that will move the crystal along a straight-line path in increments of fractions of a micron—less than one-ten-thousandth of an inch. The conventional technology for moving an object along a linear track is a rotary-driven actuator—essentially a finely threaded screw. But its performance is limited by friction, wear, jitter, backlash, and temperature effects. These problems can't be fixed with feedback loops and smart control—the underlying mechanical technology is too bulky, slow, and frictional to be perfected that way, or any way. No mechanical system can be. The answer has been found, instead, in the direct-drive motor. Such a motor is capable of providing position control of submicron precision—at least an order of magnitude better than conventional solutions. There are hundreds of such machines emerging to power the new manufactories, with silicon emerging to eclipse cement and steel as the material at the center of high-value design, assembly, and construction.

The technologies of the chip fab are also being enlisted to fabricate micro electro-mechanical systems (MEMS) that monitor motion and pressure, move fluids, and perform countless other functions within micron-scale structures a thousand times (or more) smaller than the devices they replace. In the medical and public health arenas, for example, functional MEMS are now being built to measure molecular weights and analyze molecular structure. Chip fabs can build these microlabs on existing assembly lines; with the new tools of power now at hand, the devices are in many respects easier to fabricate than logic chips, because their structures are a lot bigger than the gates doped and etched onto microprocessors. These microlabs weigh biological particles by depositing them on microscopic surfaces that vibrate at radio frequencies—heavier particles cause larger changes in the pitch of the chip-scale tuning fork. Or they analyze chemical structure by projecting tiny beams of finely tuned radiant heat or laser light, and then detecting what the molecules thus excited send back. Or they perform full chemical assays using tubes, pumps, and reactor vessels built into the same silicon chips. By such means the new devices probe physical and chemical signatures just as full-scale laboratories do, but cheaply and automatically. Experts once scoffed at the notion of putting a computer on every desk. Before long, full-blown medical laboratories will be as small, cheap, and ubiquitous as microchips.

Comparable technologies are coming to every factory. The newest industrial robots—which are, by and large, complex configurations of electric servo motors—come packed with sensors, giving them tremendous flexibility, and they can now be instantly reconfigured to perform new tasks through software alone—a dramatic advance over previous systems that required hours of manual rewiring. Digital power, digital logic, programmability, and open standards transform highly specialized, product-specific manufacturing machines into general-purpose, atom-crunching material processors.

The new world of dexterous robots represents a huge increase in capacity for digital control. The number of sensors and logic chips on the premises doubles and redoubles again and again, in tandem with the spread of power chips. Every power chip is effectively programmable, by way of the logic chip behind it. Every logic chip can communicate in both directions, to receive real-time direction and to dispatch real-time reports. Every logic chip, along with every sensor, sends a stream of digital data on up to the higher tiers of the material-moving network. Within a decade, these lower tiers will represent more embedded microprocessors and more in-use bandwidth, than all the more visible and familiar layers higher up.

It took less than twenty years for data and telecom to make the transition from a niche sector to 10 percent of the entire economy. But nearly 30 percent of the economy remains firmly lodged in the industrial and manufacturing sector. Much of what happens in the 60 percent of the economy defined as services involves factory-like atom-moving skills, too, like cutting hair and replacing heart valves. It has taken longer to make these rest-of-the-economy layers digital, because they require much more—not just smart chips and their networks, but also an entire new generation of power chips and sensors, and analog/digital converters. The logic chips arrived in the 1980s and 1990s. The high-power power chips and the high-precision embedded sensors arrived later. The digital office came first. The digital factory comes next.

POWER AND JOBS

Suetonius tells us that the Emperor Vespasian richly rewarded the inventor of a labor-saving device that would have efficiently transported some heavy columns to Rome—but refused to use it—"you must let me feed my poor commons," he explained.[7] Not every emperor is sensitive to such perils, but the Luddites of each generation certainly are. And at first blush, the new digital factories are the Luddites' worst nightmare. The powered machines build everything, the workers build nothing; workers close at hand would indeed just add dirt and dust and are therefore banished from the premises. The jobs are lost not to the foreign worker who labors for a pittance, but to the extremely expensive but fantastically productive powered machine, which ends up laboring for even less.

The Luddites are always right—about the first half of the story. The beast of burden—followed by the farmer—did eventually go just the way the Luddites feared. A century ago, horses still provided much of our motive power; today, they have almost completely disappeared from our economic landscape. The number of Americans employed in farming began to drop rapidly in the early decades of the twentieth century, and the decline continues to this day. All the while, U.S. agricultural output spiraled upward. Breeding better solar-powered engines—crops themselves—has certainly contributed significantly to this success; the second principal factor has been the steady increase in this sector's use of powered machines. When American farmers—in ever-dwindling numbers—mount their combine harvesters to compete against farmers on the Kathiawar peninsula still riding oxen and donkeys, it is no contest at all.

And the Luddites are always wrong, too, about the second half of the story. The sons (and then daughters) of American farmers quickly found more productive and profitable work elsewhere.

Visionaries of every era repeat the monocular mistake. While Gandhi celebrated India's independence (though not the final triumph of the hand loom) in 1947, George Orwell was imagining a world in which machines had eliminated "hunger, overwork, dirt, illiteracy, and disease," but had end up under the control of a small oligarchy of "business executives, technicians, bureaucrats and soldiers." The year—1984—arrived on schedule, but somehow the oligarchy didn't, at least not in countries that mastered and embraced the technologies Orwell most feared.

While Orwell was typing, American women were buying toasters and washing machines. In short order, electrically powered appliances had taken over much of what had traditionally been "women's work" in the home. As these changes unfolded, quite serious people wondered what might be left for women to do, after we automated cooking, laundry, dishwashing, and the rest. Vacuum cleaners, washing machines, dishwashers, and large freezers arrived on cue and women were indeed laid off by the millions, from their essential but thankless drudgery in the home. They didn't stop working, however, they just started getting paid for it.

By the 1990s, the techno-dystopian Jeremy Rifkin was warning us, in *The End of Work*, that digital machines—computers, basically—would put an end to most of our jobs and leave teeming masses of discontented unemployed in their wake. Only a tiny nucleus of the techno-elite would

FIGURE 8.2   Energy and Prosperity

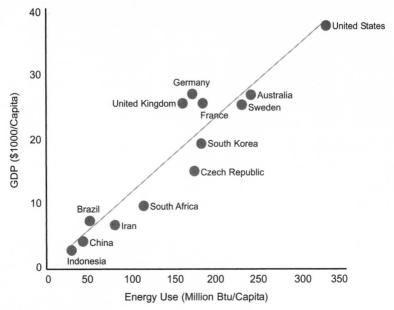

Source: CIA, *World Factbook*; BP p.l.c.

The more energy a nation uses, the richer it gets. Powered machines boost productivity, which boosts wealth.

prosper and rule. That future hasn't materialized yet, either—and all experience suggests that it won't. Throughout the twentieth century, the productivity of American factory workers was continuously raised by increasing the sophistication and use of powered engines, motors, and machines that amplified their labor. New generations of powered machines have repeatedly swept through the workplace displacing specific forms of labor when they did, but the long-term trend has been one of steady expansion in both total employment and the individual worker's income.

If we measure the work of machines by the energy used to power their labor, every person in America is now served by the equivalent of about two hundred human servants. Yet most people are still gainfully employed. Powered machines boost productivity, but decade after decade, century after century, rising productivity has somehow been matched at every step by rising employment. Today's factory worker is assembling products and using machinery that would have been unimaginable just one generation ago. Even the maligned hamburger flipper is operating electronic cookers, drink dispensers, and cash registers; her job couldn't exist at all without a vast array of refrigerated trucks, communications networks, and other gadgetry—and a transportation system that lets countless people eat an ever rising number of their meals away from home.

There is, in short, a productivity paradox very much like the efficiency paradox. The more power it puts to increasingly clever use, the more productive a workforce becomes, the more payrolls expand, and the more new jobs emerge. Wealth and the opportunities to enjoy it expand apace. The Luddites saw—and to this day see—only one half of the picture, and in the rise of the intelligent, powered machine they invariably discern the decline and fall of the most dexterous and intelligent of all productive machines, man and woman. But the decline and fall never arrive. Finding energy and harnessing power does not limit and suppress human labor—it is what keeps the worker moving forward.

9

INSATIABLE DEMAND

> The source of material civilization is developed power. If one has this developed power at hand, then a use for it will easily be found. . . . The way to liberty, the way to equality of opportunity, the way from empty phrases to actualities, lies through power; the machine is only an incident.
> —HENRY FORD (1926)[1]

As we have seen, most of the energy we consume is used to process and purify energy itself. We extract stored chemical energy from reservoirs scattered below the ground, then move it through layer upon layer of hardware—refineries, furnaces, boilers, turbines, motors, filaments, and semiconductors—and dissipate copious amounts of it at each stage, all for the purpose of extracting a trickle of distilled power at the end of the line. Somehow, the more efficient we grow at doing these inherently wasteful things, the more energy we consume. What exactly do we get from all this wasteful effort? And whatever it is we get, why do we always seek more?

SAVING TIME

Set aside pure power for a moment: what we really want is speed. And we crave speed everywhere because it saves time, the scarcest resource of

138

all. We demand faster cars, trains, and planes, faster computer and Web connections. We even demand faster televisions: according to one study, the average home uses about 5 percent of its electricity powering the instant-on circuits in TVs and other appliances,* because when we want Letterman, we want him *now*.

But in discussing time, and thus speed, we can't set aside pure power, because the two go hand in hand. It takes increasingly pure power to speed up the car, factory, computer, or connection to the Web—purer power isn't all it takes, but it's the essential starting point.† Mechanical power is purer than thermal power and concomitantly faster. The Savery steam engine of 1698 was a purely thermal device; it had no moving parts at all, other than hand-actuated valves, and was therefore very slow. James Watt didn't discover steam-powered motion, he just designed a much faster thermomechanical way of controlling it, and that made all the difference. Faster trains and cars and propeller planes weren't practical until the internal combustion engine displaced the comparatively slow steam engine. Jets weren't practical until the gas turbine displaced the comparatively slow piston engine. Electric power is faster still—much, much faster.

We value fast engines as much, and more, when they move nothing of any substance at all. Current microprocessor clocks run billions of cycles per second (gigahertz (GHz)); fiber-optic telecom links convey trillions of bits per second (terabits (Tb)). The marvel of telecommunications isn't that it lets us see things a thousand miles away—the Vikings managed that a millennium ago—but in its speed.

Or, if we insist on the importance of distance, it is the speed that delivers it. An electric heater can glow in a hearth a hundred miles away

---

* The instant-on circuit in a television draws about 8 W—a pretty big piece of the 35 W round-the-clock average for the entire device. Altogether, home appliances consume about 50 W in standby power. The circuits don't draw all that much power, but they draw it around the clock, and that's the killer. See for example, "Phantom Power," Lawrence Berkeley Laboratories, August 2001.

† Chaos can propagate at high speeds too, but we can't get anything useful out of it when it does.

FIGURE 9.1 Powering Speed

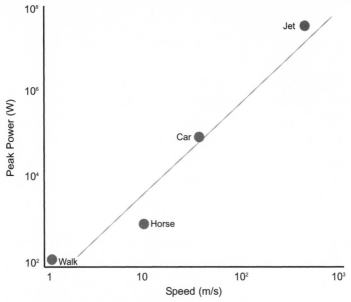

It takes increasingly pure high-density power to speed up the car, factory, computer, or connection to the Web. We consume more energy to save more time.

from the coal furnace that drives the power plant only because electricity is so compact, pure, and fast. A megaphone can't convey terabits of data through hundreds of miles of glass fiber; laser light can. It can also shoot down a missile a hundred miles away. Speed and range are corollaries, and both depend directly on the quality of power.

So we purify power to speed up everything, and by moving everything at higher speed we save time, and having saved all this time, we can then pack more miles, bits, and gourmet cuisine into our finite lives. Most of the miles we travel by car today would never have been traveled on horseback at all. Most of the miles we fly would never have been traveled in trains. Most of the bits we process would never have been run through typewriters or calculators, or viewed on paper.

We do more when we do things faster, but in energy terms, everything we do gets harder—the world wasn't designed to accommodate our craving for speed; it pushes back, and the faster we move, the harder it

FIGURE 9.2   Travel

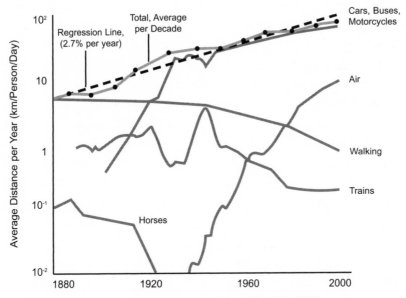

Source: Ausubel and Marchetti, "The Evolution of Transport," *The Industrial Physicist*, April 2001.

We consume more energy to extend our range. Most of the miles we travel by car today would never have been traveled on horseback; most of the miles we fly would never have been traveled in trains.

pushes. Fuel consumption rises as the cube of speed in cars (double the speed, increase fuel consumption eight-fold), hence the fuel-saving case (not to be confused with an *efficiency* case) for lower speeds. Power consumption in microprocessors rises as the square of clock speed. Fast connections to the Web consume more than five times as much power as slow ones—2 watts (W) for a dial-up phone line, 15 watts for a DSL connection or cable modem. Speeding up wireless connections increases power consumption even faster.

Speed also forces us to pile on still more power-purifying energy-consuming hardware, because speed is unstable and dangerous, and catastrophic failure can take back all the time originally saved and more. At high speed, any slight hiccup in power is likely to crash your jet or your computer—things that move very fast tend to collapse violently when even tiny things go wrong. Office computer networks thus require elaborate backup systems and uninterruptible power supplies, which generally stay hot around the clock, the always-on mirroring the instant-on-Letterman in its ceaseless demand for power. Standby diesel generators require always-on heaters to keep their oil hot to permit instant startup. Cars and jet planes have layer upon layer of redundancy in the critical parts of their power trains, because what's most dangerous isn't speed itself, but the loss of control that a loss of power entails.

And sooner or later we're bound to lose control anyway; we consume still more power protecting ourselves from speed itself. We envelop the driver in bumpers, shock absorbers, steel frames, crush zones, collapsible columns, and airbags, all of which add weight and thus require more power. To maintain both speed and safety, we end up driving vehicles with tank-like armor and monstrous engines under the hood to move them. Speed is the solution but it is also the problem, and the only way to get the benefits without the costs is to add still more power, which we do.

Creating Order

The second great scarcity of existence is order, and it takes purer power to deliver more of that, too. "Order" is a subtle concept that eludes simple definition—but however difficult it may be, this is where all the important energy accounts begin and end. Very roughly speaking, we build order whenever we choreograph motion, move things around to build new logical structures, or systematically rearrange the state of our environment. Order is logic incarnate; it is information in kinetic or material form. There is more order in a brick than clay, more in silicon than sand, and more in a silicon Pentium than a silicon crystal.

Much of the order-building power is directed at our own persons or into our homes—when power gets pure enough, we can safely embrace and gorge on it, in bacchanalian delight, and that, again, is exactly what we do. We use some 30 Quads of energy for transportation. Most of that gets channeled through engines and gear boxes and cushioned seats to deliver perhaps one Quad of kinetic energy (at the very most) to our own bodies.* One Quad is still a huge amount of energy—ten times more than we get from our food—but so long as it remains under tight control, it doesn't harm us in the least. We aren't even aware, when we're driving, of the kinetic energy that permeates every particle of our being. When humanity eventually dispatches travelers toward other stars light-years distant, in spacecraft accelerated to speeds that begin to be measured in fractions of the speed of light, we will imprint upon each individual astronaut more energy than was consumed by all of humanity in hundreds of years of its early existence.

We use about 40 Quads of energy to move electricity. About one-fourth of the total gets channeled directly into our homes, where it delivers perhaps a half Quad's worth of light, cooler air, ice in the fridge, and hot water swirling through the dishwasher. Purer power lets us move the light bulb and refrigerator motor close in, where we want them, because purer power can be transmitted cool and clean. The power plant is still hot and dirty, of course, but it's now a hundred miles away, where it doesn't bother us. However glib that observation may sound, increasing order always comes down to dumping disorder somewhere else, and the farther away the better. The second law of thermodynamics allows for no other strategy.

Perhaps 6 out of the 40 Quads are ultimately funneled into even more refined forms of electricity to power and cool computers, chip fabs, lasers, and communications channels. Only the tiniest fraction of that energy emerges as useful power at the far end, but this trickle is now so very pure that it can create order that is almost unimaginably delicate and fine. There is no such thing as coal-furnace surgery, but there is laser surgery, and while coal-fired grid power is good enough to power a light bulb, the electrically fired laser gives us back the eyes of a twenty-year-old. We teach children not to touch an electrical outlet—but we implant defibrillators to dispatch 500-volt jolts of highly ordered power to the human heart, to restart it when the body's own electrical system goes haywire.

---

*And we give it all back, of course, when we slow down.

It is extremely pure power, finally, that lets us reach down into the microcosm to build quantum structures atom by atom, layer by layer, in the semiconductor devices that define the digital age. In the heart of a chip fab, a layer of material is deposited or removed; a platter on which a wafer is mounted is rotated; and another layer is deposited or removed. Diffusion processes use thermal energy to inject atoms directly into an existing crystal; ion implantation techniques use electrical propulsion. Chemical vapor deposition directs highly reactive gases toward the surface of a crystal, where the precursor chemicals combine to deposit a new, atomic-scale crystalline layer. Focused ion-beam machines dispatch a beam of gallium ions to perform stunningly fine repairs on completed devices. Pulsed-laser deposition uses carefully shaped ultra-short laser pulses to ablate difficult materials from a source bar and deposit them in fantastically pure atomic layers on target substrates.

Huge amounts of energy go into purifying the raw materials used in erecting these atomic-scale edifices. Hundreds of tons of earth and rock are mined to extract the pounds of rare earths, lanthanides, erbium, cerium, gold, gallium, tantalum, bismuth, indium, and perovskite that serve as raw materials. Chip fabs pump air and water by the ton in a perpetual struggle to keep things clean. Intel ranks its clean rooms in terms of tens or hundreds of dust particles per cubic foot of air—far, far cleaner than a hospital's surgical theater. And within the clean rooms are still cleaner spaces where air is as unwelcome as dust. The molecular and ion beams converge on the silicon crystals in huge, gleaming steel chambers evacuated to a near-perfect vacuum by fantastically elaborate pumps. The builders of semiconductor junctions work their magic in defiance of the inexorable tendency of the universe to mix atoms up; nothing short of utter emptiness can stop that process.

SUPPRESSING CHAOS

Even as it separates and orders things in one place, energy consumption accelerates the mixing in others, and all the uncontrolled mixing is often called pollution. This mixing can be suppressed or reversed as well—but only by using still more energy. To begin with the easiest example, many householders recycle only a few simple commodities like glass, paper, and aluminum, and they do that badly if at all. Such processes can be automated—the dexterous robots that assemble the car or dishwasher can likewise be enlisted to disassemble and sort its parts when it's discarded. This won't generally reduce energy consumption, it will raise it— but it undoubtedly can reduce our need to mine new aluminum, dig up more iron ore, and harvest more trees. Most recycling as we know it today doesn't save energy either—not if we properly account for all the extra collection, processing, and manual labor in the home.

Well-ordered power also lets us detect trace pollutants far more effectively than we could before, and detection is the key to better control. The best way to monitor emissions in a tailpipe, for example, is with laser-based multispectral radar. The same technology that, in search of oil, guides a remotely operated vehicle through clouds of sediment at the bottom of the sea is engaged here at the end of the fuel cycle to track the particulates and chemical by-products created by the oil's combustion. A high-power free-electron laser, tunable from the ultraviolet to the mid-infrared bands, can likewise send megawatt (MW) micro-pulses of light through the atmosphere to identify a wide range of atmospheric aerosols and pollutants. Lower-power sensors that probe exhaust gases can be tied in to engine control systems to optimize tuning on the fly.

To reduce pollution further, we scrub and filter. We use platinum catalysts to complete the oxidation of incompletely burned fuel and nitrogen in tailpipes and reformers to strip the carbon out of fossil fuels, leaving only the hydrogen. The scrubber installed in the smokestack of the coal-fired power plant strips out sulfur—consuming 5 percent of the plant's output in the cleanup. Catalytic converters in car exhaust systems strip out nitrogen oxides—and cut fuel economy too, by using some part of the engine pressure to move the polluted gas into close enough proximity with the catalyst for the cleanup to occur. All "extraction" processes consume high-grade energy, whether they extract fuels from the disorder of the Earth's crust, recyclables from trash, or pollutants from the chaos of a car's exhaust. To contain solid nuclear wastes we literally mine the Earth a second time, to build secure, fortified repositories, much as we cocoon our children in more steel to keep them safer on the highway.

We don't yet scrub carbon dioxide out of coal plant smokestacks, but we may do that too, in time. The basic chemistry isn't difficult. Under one proposal, flue gases will be pumped through a tank with a water spray that then flows over a bed of limestone particles. The carbon dioxide dissolves in water to form carbonic acid, which then reacts with limestone—calcium carbonate—to form calcium bicarbonate. The process would effectively emulate the geochemical process of carbonate weathering which, over the millennia, takes carbon dioxide out of the atmosphere and deposits it in solid form at the bottom of the ocean.

All this will require still more high-grade energy. Chimneys use the natural buoyancy of the hot gas as a pump, to draw air through the furnace at the front end. Scrubbers of any kind lower the throughput, and carbon scrubbers would require new electrical pumps of their own, to move and mix the carbon dioxide, water, and calcium carbonate. The energy used to pump water and grind limestone would amount to about 10 percent of a power plant's output—consuming an additional 100 million tons per year of coal, at current U.S. levels of consumption.

And then we use still more high-grade energy to get rid of low-grade energy. Heat—chaos—is where ordered power ends up when the order dissipates. The more power we pump into an engine, motor, switch, microprocessor, laser, or digital radio, the more we lose as heat. And the smaller the structure, the worse the problem—more heat in less space means higher temperature, which eventually means meltdown.

If it isn't effectively dissipated, heat takes microprocessors, sensors, and engines to the grave with it. Heat is the insidious enemy of valves, seals, capacitors, and logic. A car's engine pumps water and spins fans to remove the heat produced by the engine itself. Microprocessors sprout cooling fins and fans, and air conditioning is essential in any room where more than a few microprocessors run full-time. The highest-resolution imaging systems and sensors require cryostats to keep them bathed in liquid nitrogen, otherwise they're blinded by the thermal background. Solid-state lasers must be cooled by tiny, solid-state Peltier chillers. The faster the clock and the smaller the structure, the more power we have to use to pump out the power after we pump it in.

WAGING WAR

The projection of power is the essence of war. As the great economic historian Carlo Cipolla recounts in his 1985 classic *Guns, Sails, and Empires*, technology played the key role in conquest and colonization from 1400 to 1700. Before, the Europeans had been weak, vulnerable (to the Turks particularly), and quite incapable of challenging distant empires like China, which were considerably more advanced on many technological fronts. Then Europeans mastered two key technologies of power: cannons, which allowed them to project terrifying power from a safe distance, and ships, which gave the cannons an effectively global range. "Exchanging oarsmen for sails and warriors for guns meant essentially the exchange of human energy for inanimate power." Europeans "broke down the bottleneck inherent in the use of human energy and harnessed, to their advantage, far larger quantities of power. It was then that European sails appeared aggressively on the most distant seas." The "one and unequivocal"[2] consequence of mastering gun and sail was a steadily growing European dominance worldwide. For the technologically laggard countries, "things turned progressively for the worse."

Today, half a millennium later, sailing ship and cannon have given way to an astonishing array of power-projecting military technologies. The pilot of the digitally powered airship sits hundreds of miles away, steering a comparatively tiny platform that bears digital eyes and digital artillery. Fully functional bat-sized, then butterfly-sized, autonomous vehicles have already been built. AeroVironment's electric-powered Black Widow—a 6-inch-span, fixed-wing aircraft—typifies a new family of tiny flyers, with 2-mile range and a live color video downlink. The company, along with several others, is now developing a wing-flapping, dragonfly-like Microbat that weighs half an ounce, including its camera and telecom downlink.

Like the act of destruction that it immediately precedes, the targeting of the new weapons centers on the projection of highly ordered power—not enough power to destroy, just enough to see. Harold Edgerton's stroboscopic photography illuminated nighttime targets in World War II; the magnetron and radar emerged in tandem to save London and the Atlantic convoys. Today the military extends digital power technologies to see across the entire span of the electromagnetic spectrum, from the thousand-kilometer-length radio waves used to communicate with submarines, to the millimeter waves in the 10 to 30-gigahertz bands that provide radar for fire control, to the submillimeter 30 to 90-gigahertz high-bandwidth satellite links, to the micron-scale waves in the 1,000 terahertz (THz) infrared bands used for night vision.

The historical advance of the technology of warfare can be collapsed onto a single chart, which tracks rising power density on one axis and rising speed on the other. The weapon that projects more power through a narrower window, faster and more accurately, prevails. That principle holds true at every stage in the projection of force. It holds for the technologies that project power to detect the targets at which the weapons are aimed. And for the delivery platforms—the missiles, jets, helicopters, ships, and tanks—that move weapons to the battlefield. And for the weapons—lasers, cruise missiles, electromagnetic-pulse generators, armor-piercing projectiles, guns, and explosives—that project destructive energy to its ultimate target.

The new technologies of power are increasingly being harnessed, as well, to project power not as chemical explosive or kinetic energy but as photonic pulses in the microwave and optical frequency bands. Electromagnetic-pulse weapons can destroy delicate wiring in all things electronic over huge areas—from cell phones and PCs to generators and engine ignition systems; at higher power levels they can disable human targets and punch through armor. The Army is developing 200-kilowatt (kW) solid-state lasers using commercial semiconductor technology inconceivable even a decade ago. Or pump up millimeter-wave power high enough and it can cook things, or people, or hostile microorganisms, at quite a distance. Such weapons can make anyone within several hundred meters feel like his whole body is touching a very hot light bulb, which of course impels flight. The Pentagon calls this "active denial technology." Shift the frequency a bit and go for a longer pulse, and the "denial" becomes terminal.

Overwhelmingly, the technologies of war now depend on the ubiquitous adoption of power semiconductors in the civilian sector. Large defense contractors still assemble the guidance system plus explosive in a smart bomb, and the complex mix of steel and silicon that makes up a Nimitz-class aircraft carrier. But the core components and tools that account for much of the cost and all of the astounding precision and agility of the new weapons and platforms—smart, powerful chips, together with many of the countless layers of software that make them function—are manufactured by the same companies that build microprocessors for PCs and power amplifiers for cell phones. It is the huge civilian demand for PCs, palm pilots, cell phones, high-tech cars, and smart appliances that has made these components as cheap and disposable as bullets and fuses.

Integrated circuits emerged from aerospace programs in the 1960s; gallium arsenide semiconductor amplifiers that make possible the compact, cheap cell phone were pioneered by TRW for defense purposes a decade ago. The gallium nitride and silicon carbide power chips that will land in consumer electronics a decade hence are being developed today in R&D programs funded by the military. But time and again, the technologies became affordable, even for military use, only after they made the transition into civilian sector, where manufacturing processes get scaled up for mass production.

As a result, the socioeconomic structure of the U.S. armaments industry today looks more as it did 1911, when Dwight Eisenhower entered West Point, than in January 1961, when he warned Americans, in his farewell address to the nation, to beware of the "grave implications" of the new "military-industrial complex." It takes far fewer people to fight and direct wars today than it did even a decade ago, because the speed and power of the front-line soldier has been so greatly amplified; our distant wars are now fought, once again, by a few, a happy few, a band of brothers, while the rest of us lie a-bed, watching their progress on Fox. The military-industrial complex now consists of two relatively thin book-ends on our enormous civilian high-tech economy. Military R&D programs push the leading-edge development of power semiconductors, software, and sensors, a decade or so out ahead of Intel, Motorola, or DaimlerChrysler, and then encourage the migration of successful technologies out into the civilian sector as quickly as possible. Military contractors end up buying back the same technology at mass-production prices, to embed it into every mobile platform, weapon, and projectile on the battlefield.

That the new technologies of war rarely figure in books about "energy" reflects only on the low quality of most of the discourse in the field. These technologies exist only as natural successors to the voracious technologies of power that preceded them. They are the latest chapter in the chronicles of coal and steam, gasoline and piston engines, kerosene and gas turbines, pumps, trains, cars, light bulbs, and radios.

POWERING LOGIC

As we discussed in chapter 2, digital logic is itself nothing more nor less than highly ordered, meticulously choreographed power. Our demand for power of this kind is growing exponentially. Thirty years ago there were no personal computers; today, the United States has an installed base of two per household—two-thirds of them outside the home—and many of them running day and night. A simple desktop unit consumes 10 to 100 watts even when it's idle—and considerably more when it's working hard and is connected to a large monitor, printers, and other peripherals. The humble phone line for the modem consumes 2 watts; a broadband connection, 15 watts; a PDA recharging cradle runs from 2 watts empty to 12 watts when charging. Reconfigured as a digital VCR that trolls the airwaves day and night for programs to record, the computer consumes about 30 watts day and night. To put these numbers in perspective, Americans consume about 1,500 watts, when all loads are averaged out per capita and across the twenty-four-hour day.

Logic machines require so much power because they are so dreadfully forgetful. Their capacitors and gates are extremely fast because they are extremely small, but that means their insulating layers are small too, so they're electrically leaky. All logic engines continuously degrade the quality of the power moving through them, and thus the integrity of the information that they store. The degraded power must be pumped out as waste heat—and new, ordered power must constantly be pumped in to restore informational integrity.

Building smaller gates appears to curtail losses in that less is lost per gate. Smaller gates must also be run at lower voltages, because they have thinner insulating layers between them. But with every step they take down this staircase, chip engineers find that they can run the gates faster and pack them more densely, so they do. More gates and higher speeds mean more power consumption for the device as a whole. Thus, tens to hundreds of watts, fluctuating at gigahertz speeds, have to be pumped into today's microprocessors—and the power requirements keep rising, not falling.

The only way to get that much power into that small a space is to use *higher* voltages. But the only way for such a small a gate to survive is to run at *lower* voltages. Thus, the discrete logic element heads south toward zero voltage and power, while the overall power consumption of the intelligent chip heads north toward infinite voltage and power. The faster the logic, the closer the power supply has to be packed in with it. The biggest long-term challenge facing chip engineers is how to bridge this zero-infinity discontinuity, how to push more power, faster, into larger aggregations of faster, lower-power gates. The fundamental limits imposed by the laws of physics are still far from clear. They certainly exist, however, and they are probably defined by the basic physical properties of materials.

In telecommunications, the maximum capacity—bandwidth—of a communication channel is directly related to the average signal power: the higher the ratio of the signal power to the background noise, the faster it is possible to transmit data through the channel. It requires no power at all to transmit information if one has infinite time, but anything faster requires power in direct proportion to speed. Claude Shannon worked out the rules in basic theory in 1948.[3] In actual practice, private, commercial, and military radio-frequency transmissions together consume about one percent of all our electric power; we thus burn about 10 million tons

FIGURE 9.3   Core Microprocessor Trends

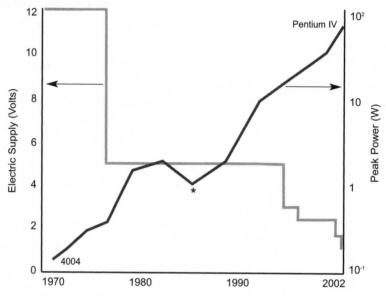

* Introduction of the CMOS transistor architecture

Source: Intel.

As gates on a chip get smaller, operating voltages must be lowered or the insulating layers break down. But as gates get smaller, more get packed on to the chip, power consumption for the chip as a whole goes up, and higher voltages are required to pump in the power.

of coal per year (and the energy equivalent in gas and uranium) to propel weightless photons through space.

No counterpart to Shannon's Law has yet been worked out for computing. For many years, physicists believed that the minimum energy required to perform a single logic operation could not fall lower than the background level of thermal noise that exists in any physical structure.* It has since been theoretically established that there is no minimum amount of energy required to perform a single logic operation, so long as the hypothetical power-free computer can run as slowly as it likes.[4] But there is nothing very profound about energy analyses that assume

---

*This quantity is directly related to temperature T and expressed as kT, where k is Boltzmann's constant.

infinite time—the second law of thermodynamics is a statistical law, and given enough time, it too can be violated. No one has yet developed a formal theoretical analysis for the minimum energy required to perform a logic operation at any specified speed. A Nobel Prize probably awaits the team of young physicists, mathematicians, and material scientists who eventually do.

The theoretical answer, when it emerges, will likely confirm what the practical world of chip engineering has already made clear: the faster you want to run a computer, the more power it will take and the sharper the power/logic schism must become. "Zero-power" will turn out to mean "infinitely slow," in computing just as it does in communication. The theory will end up showing what every real-world chip engineer has already surmised: an "infinitely fast" computer will require infinite amounts of power, delivered infinitely fast, into an infinitesimally small space.

If such a computer has ever existed, it existed some 14 billion years ago or thereabouts, within the interstices of what cosmologists call the Big Bang.

ENRICHING LIFE

As we noted in chapter 3, the word *entropy* is derived from the Greek "in-turning," an allusion to the seemingly universal drift toward chaos. But in limited spaces, for limited times, with suitable supplies of high-grade energy in hand, it is possible to push things in the opposite direction, toward order and away from chaos. We crave energy because we crave order, because order is life and chaos is death.

With highly ordered power, we can build a highly ordered material environment. Industrialized nations move vast amounts of material through their economies—an estimated 35 tons of material per capita, per year.* Perhaps, at some point, we won't care to expand our ton-moving

*For the United States, 25 metric tons/yr/capita nonfuel materials (Jocelyn Kaiser, "Turning Engineers Into Resource Accountants," *Science* 285, no. 5428 (30 July 1999): 685–686; in addition to approximately 10 metric tons/yr/capita in fuels.

enterprises any further; we will have enough wood, steel, and cement for our homes and factories, and enough tarmac for our cars. Even if we don't "run out of room" here on the surface, we may well reach the point where we simply value untamed nature as much as we value high-rises and highways. But as Richard Feynman pointed out in a lecture delivered in 1956 (to which we return in chapter 12), there is plenty of "room at the bottom," in the atomic junctions where we are now building the structures of the digital age.

However crowded things may get up here, we will continue to pour energy into reordering the "staggeringly small spaces" Feynman alluded to—first to build atomic-scale junctions, then to power the quantum devices so that they can sense, calculate, and communicate for us. Pound for pound, it takes much more energy to build a Pentium or a laser diode than it does to launch a Space Shuttle. About 1.5 percent of all the electricity used in the United States goes into separating aluminum from oxygen, because you can't make a soda can with the aluminum oxide, but you can make it with pure aluminum.[5] We may eventually satisfy our appetite for soda cans, but we will never stop finding new uses for one-micron aluminum wires to conduct electricity on the surface of microprocessors, at least not until we master some even finer process that uses a better material. We will never stop wanting more order at the atomic scale of things, or needing more highly ordered power to create it, and still more power to use those atomic-scale structures to create still more order.

Once built, smaller structures consume far more energy, too, in proportion to their size. And we build far more of them. LEDs will be installed in far more places than incandescent light bulbs, because they are cool and compact enough to be mounted on clothing or monster outdoor TV displays. We build far more cars than trucks or locomotives, and far more PlayStations than ENIACS, and in the aggregate, the smaller systems end up consuming more power than the bigger ones.

Well-ordered power isn't an end, it's a means toward human ends— *all* of them. If we want to embrace all those ends in a single word, the emphasis has to be on order, not energy. We don't have a precise definition for that word either, but at least we recognize how abstract the concept is, we don't pretend to understand it well, and we don't spend time urging others to curtail their "consumption of order" or declare that less order—which is to say more disorder—is the only way to save the planet. And who supposes that we will someday satiate our hunger for information, the ultimate form of order, or for more safety, another form, or for better health or longer life, the span of biochemically structured order that defines our own existence?

Our power to create and maintain order is what feeds, moves, informs, and entertains us. It's what we engage when we travel across the oceans, or excise a cancerous tumor from deep inside our bodies. It's what we turn on when we beam high-power X-rays through luggage in airports, and what we rely on when we pursue terrorists into the deep recesses of distant caves. It's what lets us dispatch rockets to explore other planets, write books, compose songs, or film movies. We will never stop wanting more logic, more memory, more vision, more range—all of which depend on high-grade energy—because we are built to want more of these things, an unlimited more.

10

SAVING THE PLANET
WITH COAL AND URANIUM

All the forests will be gone [by 1993]. Lumber will be so scarce that stone, iron, brick, slag, etc., will be largely used in the construction of houses. As a result, fires will be almost unheard of, and insurance companies will go out of business.
—JOHN HABBERTON, "OF WOMEN, LITERATURE, TEMPERANCE, MARRIAGE, ETC."(1893)[1]

So HIGH-GRADE ENERGY buys us order, which we crave insatiably. But in the grand scheme of things, the pursuit of order is a loser's game; the first law of thermodynamics says you can't win in the energy racket, the second decrees that when playing for order you can't even break even. For every unit of new order you rake in on the green felt, you lose a unit and then some from a bank account a hundred miles away.

Most of the time we don't much care about those distant losses, and we shouldn't. Most of the new disorder created when we burn coal or gasoline is just plain heat, which we cheerfully dump into the air and forget about. Thermal pollution can be a real issue when, for example, it disrupts the ecology of a river. But most of the time it simply isn't—the amount of heat we release is minuscule compared with the amount that

156

cascades down from sun to Earth by day, and then radiates back into the depths of the cosmos by night.

More bothersome are the nitrogen-oxygen (NOx) compounds created in every air-based process of high-temperature combustion, and the sulfur dioxide (acid rain) and particulates that all lower-grade fuels release when they burn—wood being by far the worst, per unit of useful energy delivered, though the poorest grades of coal aren't much better. As noted in the previous chapter, however, these problems are quite readily addressable with today's technology and more-energy solutions. New cerium catalytic filters can all but eliminate particulate emissions from diesel engines. Tailpipe emissions from Honda's new ultra-clean internal combustion engine are cleaner than the ambient air on the Santa Monica freeway. Burn even more fuel, dump even more waste heat, and it's reasonably easy to scrub conventional pollutants from the flue gases of power plants and the tailpipes of cars.

Carbon dioxide is another matter. Weaning ourselves from hydrocarbons themselves is something that we will undoubtedly do someday, but no time soon. Scrubbing out the carbon dioxide at the smokestack (though not the tailpipe) is technically feasible, but given the gargantuan amounts of carbon at issue, this would require huge additional capital investment and concomitantly large increases in the consumption of fuel.

Would it be worth it? That depends entirely on how seriously we take the claim that human carbon emissions are changing the global climate. For the foreseeable future, the best (and only practical) policy for limiting the buildup of carbon dioxide in the air is to burn more hydrocarbons—not fewer. And then, more uranium.

CARBON CHAOS

Because it now figures so centrally in the policy debates, let's use carbon itself as our quantum metric of "disorder." This is *not* thermodynamically rigorous—far from it—but it will do as a surrogate for discussion, particularly if the global warming models are right. Highly imperfect though the carbon metric is, it does give us one systematic way to line up benefits and costs. One hour of the order that we call 100-watt (W) light costs us, on average, 0.05 pounds of atmospheric-carbon chaos. One bucket of ice from the refrigerator, 0.3 pounds. One average hour in a car, 5 pounds. The pounds certainly do add up. In the units commonly used in the scientific literature, fossil fuel combustion releases about 1.8 billion metric tons of carbon per year into the North American air. Worldwide, humans emit roughly 6.5 by burning fossil fuels, and another 2 through deforestation.[2]

These are big numbers—but even so, they must be viewed in perspective. Plants—the green kind, not electric power plants—exhale about 59 billion metric tons of carbon (in the form of carbon dioxide) a year, and absorb roughly 120 billion in photosynthesis.[3] Soil organisms, digesting the dead plants on which they live, emit 59 billion. A net of about 26 billion physically diffuse into the atmosphere out of the oceans, and about 28 billion diffuse back in. In short, green plants and "carbon weathering," both powered by the sun, continuously establish new carbonaceous order. Pretty much all the rest of life promotes carbon chaos. The guess-timated bottom line: chaos is currently gaining ground, at a rate of 3 billion metric tons of carbon per year. Without human combustion of fossil fuels, the order might be gaining at about 4 per year.

On these carbon-based order/chaos books of account we are dealing with small differences between large and uncertain numbers. We do know that concentrations of atmospheric carbon dioxide rose about 20 percent in the past century—but we also know that concentrations have varied substantially in the past, long before fossil fuels entered the picture. Carbon dioxide levels were only half as high some 50,000 years ago, but they were almost as high as today 150,000 years ago. Eight hundred million years ago the Earth's air was mostly carbon dioxide. Green plants evolved and flourished in profusion and sucked up most of it. Some of the plants sank into swamps, and then sank deeper. Hence the fossil fuels that we now burn in such quantities.

The fear is that if we dig up and burn all the fossilized plants of the Carboniferous period, we can expect to re-create the atmosphere of that period too—a carbon-rich hothouse. Climate models assume carbon

FIGURE 10.1   Global Carbon Flux
(Billion Metric Tons Carbon)

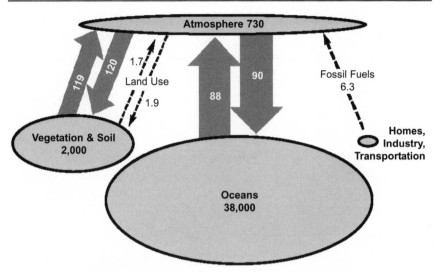

Source: EIA, International Energy Review 2002 (data are from 1999).

Fossil-fuel combustion and deforestation release carbon into the atmosphere. Human agriculture and forest regeneration remove it. Much larger carbon fluxes are propelled by plant and animal life in the rest of the biosphere, and by the weather over the oceans.

dioxide triggers a first little bit of warming. On its own, the effects are still inconsequential—carbon dioxide alone, in the quantities we add, does not act as a very effective atmospheric blanket to block the shedding of heat (and with it, disorder) from the surface of the Earth at night. But water vapor might amplify the impact significantly. Warmer air holds more vapor, which blankets the planet a bit more, which warms the air still more, which holds more vapor, and now the Earth becomes a runaway greenhouse. Or so a good number of the climate models suggest. There is much uncertainty to these models—far more than is acknowledged in most accounts. But the mere possibility that we might be changing our *global* environment is indeed worrisome.

How, then, can anyone responsibly favor the burning of more hydrocarbons? The short answer is that for most people, the only practical alternative today is to burn carbohydrates, and that's much worse.

America the Beautiful Carbon Sink

Round the clock and around the year, the sun delivers to the United States an average energy flux of roughly 180 watts per square meter. And humanity has certainly found ways to capture some of that energy, albeit not yet very much of it. Worldwide, wild plants currently convert about 0.35 watts per square meter (W/m^2) of that into stored energy, a dreadful 1:500 energy conversion efficiency. But with the help of a horse, mule, buffalo, or ox, it isn't too difficult to transform the solar energy thus captured into horsepower. That's how most of the world still gets almost all of its energy—from carbohydrates. We Americans once survived on a high-carb energy diet ourselves. In 1840, it required 6,000 cords of wood to produce 1,000 tons of iron; an iron producer harvested 1,000 acres of timber a year to fire a single furnace.[4] As late as 1910, as noted in chapter 1, some 27 percent of all U.S. farmland was devoted to feeding horses used for transportation.

Feeding the organic transportation system of 1910 thus required enormous amounts of land—far, far more than we have since seized for oil pipelines, refineries, and wells.* When Europeans first arrived on the continent the contiguous forty-eight states had about 1,045 million acres of forest. That area shrank steadily to a low of about 750 million acres in 1920.[5]

We have been restoring forest ever since. Exactly how fast is hard to pin down: the continent is large, most of the land is privately owned, and the definitional debates rage about when regrowth reaches the point of establishing new "forest." But all analyses show more, not less, forest—America's forest cover today is somewhere between 20 million and 80 million acres higher than it was in 1920.[6] About 9 million acres have been reforested since 1987 alone.[7] Trees have been replanted, in recent years, at a rate of some 3 million acres per year.[8] We're adding new lumber-

---

*1910 cropland at 306,000 million acres: 27 percent = 83 million acres. Total 16 million acres for current land devoted to all energy production, transportation, and conversion. Adding in 27 million acres for all roads and highways yields 43 million acres (See Endnotes #6, 7, and 14 in Chapter 10).

quality trees 30 percent faster than we're harvesting them.[9] For the first time in history, a Western nation has halted, and then reversed, the decline of its woodlands. Within a generation, if current trends continue, America could return to levels of forestation last seen by the Pilgrims.

These numbers, though wrenchingly at odds with common notions of what just must be true, are perhaps easier to grasp and accept when stated in terms of how the average family has used the land, yesterday and today. A century and a half ago, a pioneering American family lived off 40 acres and a mule. The family lived entirely off the land, and to do so, cut down trees for crops, pasture, and fuel—just as families still do today in the Amazon basin and much of the rest of the world. Since 1920, however, the North American family has returned at least one and perhaps 2 acres of the homestead to forest. It doesn't need them anymore. Now, it's digging up its energy, in much more concentrated form, from below the surface.

Per acre of land used, agricultural productivity at least tripled in the twentieth century, in large part because so much less land is now required to power the plow. The pioneer farmer got his horsepower from his horse, which required 2 acres of pasture to feed. Better crop strains have played a key role too, along with agricultural chemicals, synthesized with copious amounts of oil. Better railways and highways, and the fossil fuels that power them, have allowed us to trade inferior farmland in New England for better land on the prairies. Highly mechanized, energy-intensive agriculture has done the rest. Overall, roughly 40 million more acres of cropland were harvested seventy years ago than are harvested today.[10]

How has this fundamental change altered our carbon books of account? At today's level of population, an American family of four can lay claim to only 30 acres of the continent, if we imagine the entire land mass evenly divided up on a per capita basis. Roughly one acre of the allotment goes for home, office, factory, road, and highway. Six acres are farmland; 8 are range for livestock; 15 are grassland, forest, mountain, and desert. Instead of harvesting carbohydrates from cleared land, the modern family digs up 24 metric tons of carbon a year, as coal, oil, and gas, and releases it into the air as carbon dioxide.

FIGURE 10.2   Total U.S. Land Use per Capita

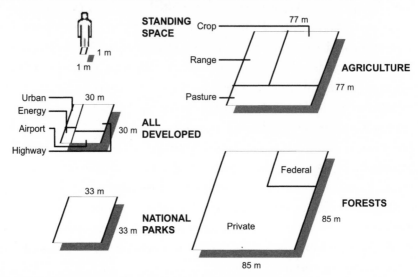

Source: Peter Huber and Mark Mills, "From Carbohydrates to Hydrocarbons," *Grenzen ökonomischen Denkens: Auf den Spuren einer dominanten Logik*, ed. Hans A. Wüthrich et al. (Gabler Press, 2001), p. 151. Data for contiguous U.S.

A century ago, a pioneering American family required 40 acres and a mule. Today, allocated per capita, the average American uses far less—about 2 acres in total for dwelling, roads, farm, range, and energy supplies.

Twenty-four tons seems like a lot. But spread over 30 acres that's about 6 ounces of carbon per square yard, or a film averaging about two-thousandths of an inch thick over the entire estate. And in North America, at least, various processes do indeed seem to be depositing that much back again, and even a bit more. Today, North America as a whole is, apparently, a carbon *sink*.

As noted earlier, fossil fuels burned on the continent release about 1.6 billion metric tons of carbon per year into the air. Prevailing winds blow from west to east. This means carbon dioxide concentrations should be 300 parts per billion higher in the North Atlantic than in the North Pacific. But in fact they're about 300 parts per billion lower. As best these things can be measured directly, America's terrestrial uptake of carbon—the amount moving down into the surface rather than up into the air—runs about 1.7 billion metric tons per year, just ahead of the amount emitted by the combustion of fossil fuels. The numbers were set out in a stunning if little publicized article published in an October 1998 issue of *Science*.*

Carbon-sink skeptics say they don't see enough new trees to account for the drop. But then, global warming skeptics say they don't see enough human carbon emissions to account for rising temperatures. The weight of the evidence indicates both a warming planet and a huge North American carbon sink. The carbon-sink numbers are, if anything, the more reliable, because they require only direct measurement today, not estimates of conditions a century ago.

And if we can't precisely explain where all the carbon is sinking, it's because it's hard to track deposits that average two-thousandths of an inch over a vast continent. Many forest inventories count only "lumber quality" trunks, ignoring younger trees and grassland. New forests mean new layers of carbon-rich soil, which are almost impossible to inventory accurately. New soil means more silt in rivers, which dump carbon into the ocean. The total forest ecosystem in the United States holds an estimated 52 billion metric tons of carbon.[11] A net growth rate of 3 percent a year is enough to consume *all* carbon emissions of the U.S. economy.[12] Either in forests themselves or on surrounding grasslands and farms, that is about the net growth rate we seem to have. The carbon chaos we create in burning fossil fuels appears to be offset, and then some, by the carbon order we create by giving back land to trees.

* Carbon Budget	Fossil Emissions (pentagrams/year)	Terrestrial Uptake (estimated range)
North America	1.6	1.6 to 1.7
Eurasia and North Africa	3.6	−0.4 to 0.5
Tropics and Southern Hemisphere	0.7	−1.1 to 0.9
Total	5.9	1.1 to 2.2

S. Fan et al., "A Large Terrestrial Carbon Sink in North America Implied by Atmospheric and Oceanic Carbon Dioxide Data and Models," *Science* 282, no. 5388 (16 October 1998): 442–446. Data are from 1988–1992.

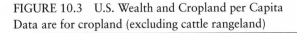

FIGURE 10.3   U.S. Wealth and Cropland per Capita
Data are for cropland (excluding cattle rangeland)

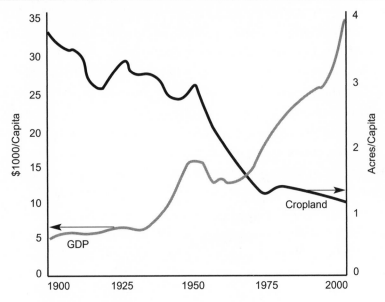

Source: U.S. Census Bureau, *Historical Statistics of the United States Colonial Times to 1970*.

Per acre of land used, agricultural productivity at least tripled in the twentieth century, in large part because so much less land is now required to power the plow. A horse required 2 acres of pasture to meet its energy needs; the oil wells that fuel a tractor occupy much less space.

CARBOHYDRATES VERSUS HYDROCARBONS

What would it take to reverse course and return today's America to a carbohydrate-fueled economy? Some people propose just that—solar energy that's literally green. "Tilling, not drilling," they advocate. "Biology, not geology. Living carbon, not dead carbon. Vegetables, not minerals." A great shift back, in short, to what a group called the Institute for Local Self-Reliance extols as the "carbohydrate economy." Which is to say: agriculture. Suitably enhanced, perhaps, to raise yields. "We need to go from black gold to green gold," declares the director of biotechnology development at DuPont.[13] Farmers certainly love the idea, and their votes have produced decades of federal subsidies for corn-based ethanol fuel.

Yet however preposterous the idea may sound, drilling for oil and building an SUV-grade highway system uses 10 times *less* land, per mile and per useful pound moved, than growing food to fuel a bicyclist. The SUV starts out 300 times worse than the bike—because it moves 30 times the weight of its driver in steel, and because it needs 10 times more roadway per useful pound moved. In terms of land surface occupied to extract and deliver the energy used, however, crude oil is at least 1,000 times more frugal than grain. And a car engine, and the refining and distribution systems behind it, are about twice as efficient in converting crude oil to locomotion than the grain-bread-muscle systems that stand between plants and the pedals on a bike; make that 16 times as efficient if the biker favors meat.

Or to turn the numbers around, 2 acres of top-notch timber-growing forest can yield a sustainable 40 tons per year of wood biomass, from which you can extract the liquid fuel equivalent of 130 gallons of gasoline, which will then propel an average car about 3,000 miles. Today, by comparison, the average American uses about 2 acres of land for *all* dwelling, roads, farm, range, and energy—the whole lot.[14] Plowing for carbohydrates adds more carbon to the air than mining for coal or drilling for oil, because the solar-carbohydrate refineries—farms—require such huge amounts of cleared land. Which means that the carbohydrate-fueled stomach is a whole lot worse for the atmosphere than the hydrocarbon-fueled motor that has replaced it.

However unlikely the numbers may sound, they are easy to verify. They are, indeed, almost obvious once one gets used to seeing plants, cows, horses, and bikers for what they are: land-hungry, territory-expanding automatons, optimized for survival and reproduction, not for supplying power to others. To be sure, a bike or a horse *looks* incomparably greener than an SUV or eighteen-wheel truck, but only because we think of cornfield and pasture as "natural."

People everywhere grasp that energy is the key to survival and prosperity, and until they can satisfy their insatiable demand for energy in other ways, they satisfy it by occupying more land. Only very recently, and only in a few countries, have people discovered how to sever the link between wealth and land. Land-poor Europe leveled most of its forests centuries ago and is now preoccupied with protecting the cow pastures of reactionary farmers. Much of the developing world still depends largely on agriculture. But—thanks to fossil fuels—North America has reversed direction.

THE SOLAR ROAD TO ENVIRONMENTAL RUIN

Growing grass in a pasture isn't the only way to transform solar energy into horsepower. For efficient, reliable, round-the-clock capture of solar energy, no other technology yet comes close to the one incorporated in the Hoover Dam, which counts on the sun to lift the water that spins its turbines. But like ethanol farms, dams may cost you a forest or two, upstream.

Selenium-doped silicon wafers mounted in glass or plastic can currently capture about 30 watts per square meter on a round-the-clock average in the United States—making photovoltaics (PV) about forty times better than the typical leaf of a green plant, and over ten times better than an intensively cultivated cornfield. But New York consumes 55 W/m^2 of energy. So to power New York with PV, you would have to cover every square inch of the city's horizontal surface with wafers—and then extend the PV sprawl over at least twice that area again.* A Pentium 4 microprocessor consumes about 20 watts per square *centimeter*—which is to say, ten thousand times more, in power density terms, than a PV solar cell can generate—and semiconductor-grade silicon for digital circuits is currently being shipped ten times faster than silicon for PV cells.

The fuels we rely on currently, by contrast, are well matched to the modern city's intensely concentrated demand for power. A coal mine yields something like 5,000 W/m^2 of land, an oil field double that, and a uranium mine, together with enrichment facilities, at least a hundred times more. Environmentally speaking, conventional fuels get steadily worse from there on out—for every acre used in extracting them, we

---

*Our numbers here are extremely generous on the solar side—we have made no allowance in our calculations for large areas required by any practical solar system for the mounting infrastructure, storage, and transmission.

use another 2 acres or so for track, pipeline, and power line to transport or transmit.[15] But refineries are needed to transform corn into ethanol, too, and generating any significant share of our electricity with PV cells would require more grid, not less, because large solar arrays must sprawl over such wide areas.* Rooftops do offer some spare real estate for solar capture, but nowhere near enough. Desert sands and rocky plateaus offer some additional ecologically dead surface but usually at great distances from where people live—which means using up more (and better) land hauling the energy back to the people.

Despite decades of subsidy and government promotion, "renewables" (other than conventional hydro) now generate barely 0.7 percent of our electricity.[16] Regulators have forced electric utilities to buy renewables— at an average price about three times higher than they pay for conventional sources of supply. Add in the forest industry's on-site burning of waste wood, and renewables contribute perhaps 3 percent of all the energy we consume.

No conceivable mix of solar, biomass, or wind technology could meet even half our current energy demand without (at the very least) doubling the human footprint on the surface of the continent. Humanity burns 345 Quads of fossil fuel energy per year and a good bit more wood and dung on top of that. All the plants on the surface of the planet capture only an estimated 2,000 Quads of energy through photosynthesis, and that energy is already being used—to grow wilderness.

GREEN NUKES

The most cogent objection to the reforestation defense of fossil fuels is that it claims current credit for simply undoing the environmental

---

*In the calculations summarized here, however, we haven't even bothered to charge a transportation overhead to "renewable" alternatives, while we did make full allowance for the rights of way for electric power transmission lines, gas pipelines, refineries, and so forth required by the conventional fuels. The fossil fuel economy ends up at least ten to thirty times more land-frugal than solar-carbohydrate economy every time, under any plausible mix of technologies.

depredations of a century ago. All the new growth of forests today is possible only because so many trees were leveled so ruthlessly by our great-grandparents. Decades hence, vast, denuded tracts of rainforest may emerge as a major new carbon sink too, as countries like Brazil complete their transition from a carbohydrate to a hydrocarbon economy. By burning fossil fuels, however, we're returning to the atmosphere carbon that hasn't been there for hundreds of millions of years.

The one demonstrably practical technology that could decisively shift U.S. carbon emissions in the near term would displace coal with uranium. Electricity is already the ascendant fuel of the digital age in every sector but transportation, and as discussed in chapter 5, even that schism is likely to disappear in the next few decades. Both coal and uranium are already heavily used to generate electricity, so we know that choosing between the two is perfectly feasible.

Whether uranium will in fact play a role in rebalancing the carbon books will depend not on those on the right wing of this political debate, who have already had the audacity to propose a nuclear revival, but on who now speaks for the greens. Is it the reflexive antinuclear ideologues? Or a more rational new center, to be led, perhaps, by the National Wildlife Federation, the Audubon Society, and other serious students of climate models and global warming?

What landed the greens of both camps in their present quandary was their joint pessimism about nuclear power two decades ago. The anything-but camp had been agitating against nukes long before the March 1979 accident at Three Mile Island. But after that accident this camp had a real meltdown to rally around. That was it, politically speaking—there would be no new nuclear plants built in the United States. The plan was to invest, instead, in efficiency and renewables, the energy technologies of the future.

What we in fact did in the twenty-plus years since was find ways to burn an additional 400 million tons of coal a year. Electric efficiency rose sharply in motors, lights, and refrigerators everywhere, but total consumption of electricity rose equally fast, so we burned more coal. We burned more natural gas, the fossil fuel favored by the greens, but we also burned more coal. We even burned more uranium, by running existing nuclear plants more hours per year, but on top of that, we also burned more coal.

What will we do in the next twenty years? Several thousand additional reactor-years of statistics since Three Mile Island testify to the safety of nuclear power itself. Its wastes do not present any serious engineering problem—uranium is such an energy-rich fuel that the actual volume of waste is comparatively tiny; it is easily converted into chemically stable forms, which are easily deposited in deep geological formations that have been stable for tens of millions of years. Operating nuclear power plants are said to be unduly vulnerable to terrorist attack, but again, because they produce so much power in so little space, the plants are easy to secure—it is just matter of erecting more steel and pouring more concrete, and then still more. The U.S. Navy sails nuclear reactors around the globe in steel ships; protecting land-based civilian plants presents a comparatively trivial engineering challenge. All the numbers, and a solid consensus in the technical community, strongly reinforce the projections made two decades ago: it's extremely unlikely that there will ever be a serious release of nuclear materials from a U.S. reactor.

Are any influential segments of the green community ready yet to trust these analyses? They should be. They certainly have come around to trusting complicated, long-term computer projections of another kind. Many greens are now quite certain they have a good grip on the likely trajectory of the planet's climate over the next hundred years. And what their climate models tell them is that if we keep on burning fossil fuels at current rates, there will be meltdown on a much larger scale than Chernobyl's, beginning with the polar ice caps. Saving an extra 400 million tons of coal here or there—roughly the amount of carbon that the United States would have to stop burning to comply with the Kyoto Treaty today—would make quite a difference, we're told.

Whatever they may believe about global warming, it's time now for all serious greens, left or right, to face up to three fundamental facts.

First, an economic fact. Demand for electricity has been rising without interruption since Edison invented the light bulb over a century ago. Short of some massive economic convulsion that drastically shrinks the economy, it will go on rising. Total U.S. electricity consumption will increase another 20 to 30 percent, at least, over the next ten years. Economic growth marches hand in hand with increased consumption of electricity—always, everywhere, without significant exception in the annals of modern industrial history.

Second, a political fact. Neither Democrats nor Republicans will let the grid go cold. Not even if that means burning yet another additional 400 million more tons of coal. Not even if that means, in turn, melting the ice caps and putting much of Bangladesh under water. No governor or president aspires to become the next chief executive recalled from office when the lights go out.

Third, a technological fact. Coal, uranium, and gas plants generate gargantuan amounts of power in very small amounts of space, which means they really can and do get built within reach of the population centers that need the power. Sun and wind come nowhere close. Earnest though they are, the people who maintain otherwise are the people who brought us 400 million more tons of coal a year.

For at least the first decade of this new century, almost all new demand for electricity will be met with fossil fuels. By coal, because it represents half the installed base, and therefore half the opportunity to expand output at the margin. And by smaller-scale gas- and oil-fired units, because new jet-engine gas turbines can be deployed much faster than larger plants, and because greens dislike them the least. The next five years are set; all we can usefully discuss now is what will come after. Will it be still more fossil fuel, a good half (or more) of it coal? Or more uranium?

"Neither," the most passionate greens will respond. And from West Virginia to Wyoming, coal miners will quietly cheer them on. "Neither" has been the official green line since 1980, when Big Coal was 400 million tons a year poorer than it is today. What more serious greens will reply remains to be seen. What they ought to do is part company with Hollywood and reach some sensible political accommodation with the nuclear industry in case their global warming projections turn out to be right.

Meanwhile, conventional fuels advance, and the forests of North America expand apace. To be sure, our carbon books would be even more solidly in the black if we could get all the new trees while burning less coal and oil too. But the practical and political fact is that the two trends are not severable, or at least weren't until the development of practical fission reactors. We can do better still on our carbon books of account, and we will. But what has already been achieved in transitioning our economy to coal, oil, and uranium is fantastically good. It should be celebrated as the environmental triumph that it is.

11

INFINITE SUPPLY

The most signal service that the steam-engine has rendered to England is undoubtedly the revival of the working of the coalmines, which had declined, and threatened to cease entirely, in consequence of the continually increasing difficulty of drainage, and of raising the coal . . . It may be said that coal-mining has increased tenfold in England since the invention of the steam-engine.

—SADI CARNOT, *REFLECTIONS ON THE MOTIVE POWER OF HEAT AND ON MACHINES FITTED TO DEVELOP THAT POWER* (1824)[1]

A PAPER PUBLISHED in *Science* in 1981 predicted that by the year 2000 it would require more than a barrel of oil's worth of energy to extract a new barrel of oil from a U.S. well.[2] Returns would go negative, and that would be that for the domestic oil economy. It didn't happen. The projection assumed we'd be drilling a lot more dry holes as years passed and supplies receded. But in fact we're drilling fewer—the hit rate is six times better today than it was in the early 1980s.

Once located, oil is now extracted far more efficiently, too—in the United States and elsewhere, as well. Operating as much as 5 miles below the surface of the sea, a remotely operated vehicle navigates, illuminates, senses, and searches by means of fiber-optic gyroscopes, ground-penetrating sonar, acoustic imaging systems, low-light digital cameras,

FIGURE 11.1   U.S. Energy Production and Cost

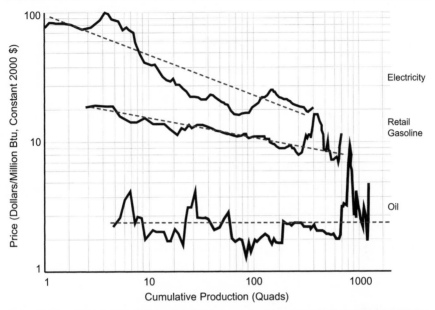

Source: EIA, *Annual Energy Review 2003*; American Petroleum Institute; John Fisher, *Energy Crises in Perspective* (Wiley, 1974).

Continuous improvement in the energy-consuming technologies that find and retrieve energy has kept energy prices stable—or falling—over the long term. 1 million Btu of electricity is a 60 W light bulb running for 6 months; 1 million Btu of gasoline is about 8.1 gallons, and 1 million Btu of crude oil is about 7.1 gallons.

LEDs, and high-power scanning lasers. The vehicle finds and maps deep-water oil fields, inspects equipment, moves and connects gears, pulls and connects pipes, and actuates valves. It digs trenches and buries cables, aligns drills with bore holes, and performs maintenance.

The continuous improvement in the energy-consuming technologies that find and retrieve energy is what accounts for the extraordinary long-term trend evident in Figure 11.1, which we included in chapter 1 (Figure 1.9) and reproduce again here. Oil extracted today from beneath 2 miles of water and 4 miles of vertical rock, with 6 additional miles of horizontal drilling beyond that, costs less than the 60-foot oil Colonel Drake was extracting a century ago and about the same as one-mile oil cost in 1980. The pessimistic view, often expressed, is that we have to work harder and harder to find and extract our energy. And so we do. Happily, however, we have more and more highly ordered power at hand to do this work—very intelligently—for us. Energy begets more energy: this is the last and greatest heresy of all, and the most important. The more energy you have, the more you get.

RISING RESERVES

Aristotle coined the word *energy*—meaning "at work"—to account for the obvious fact that it takes something to keep something else going.[3] Many intelligent people were quite surprised when Isaac Newton articulated his first law of motion, the one that says (roughly) that a moving ball or bus will keep right on moving, until some force intervenes to slow it down. But then, as Aristotle intuitively recognized, some force—friction, most commonly—always *does* intervene. In the real world here on Earth, balls and buses *do* always seem to run down. Perpetual motion is impossible. Every thermodynamics textbook still says so today.

But unlike Aristotle, the modern texts say that only about *closed* systems, which exchange no material or energy with their environment. No really interesting systems are closed, they're all open. Chaos—entropy—always increases in a completely closed system; closed systems spiral down toward equilibrium and stasis. Open systems—buses that can refuel at gas pumps, and all other systems that can pull in high-grade energy from their environment and dump low-grade heat back into it—can run up, too, if they get lucky. And in the real world that seems to happen too, all across the face of our open planet.

Starting from nothing 4 billion years ago, life somehow contrived to capture high-grade energy from here and there and used it to assemble more life—so successfully, in fact, that life, a very complex form of order, now covers the planet.* A second chain reaction of rising order got

---

*Inside the human body, the key fuel on which all the rest of our cellular chemistry depends is adenosine triphosphate—every day, our bodies synthesize and break down about 3 grams of ATP for every gram of dry body weight. Four

started with the dawn of agriculture, about eight thousand years ago. Human societies began selectively planting and breeding crops to capture solar energy systematically, and they used the expanding supplies of energy mainly to breed more people, who planted more crops. Humanity's total energy consumption doubled about every five to ten centuries thereafter, in step with the (slowly) rising population.

James Watt launched a third rising wave of energetic ordering when he radically improved an engine that could transform heat from coal into motive power—which could then be used to mine still more coal. In short order, humanity was harvesting billions of tons of plants that had grown 65 to 360 million years ago. Energy consumption and population have risen geometrically ever since.

Today, as already noted, humanity consumes 345 Quads per year of fossil fuel—which is widely supposed to be a huge amount of energy. Thus, the inevitable exhaustion of fossil fuels has been vehemently predicted since the 1970s, and somewhat less vehemently since at least the 1880s*—just as the inevitable exhaustion of food has been predicted since the 1790s, the time of Malthus. But all such predictions center on what today's technology, driven by today's forms of power, makes reasonably accessible.

No one seriously disputes that with better technology, and better power, we could retrieve far more. We already know where to find centuries' worth of coal—global deposits hold 200,000 Quads. Oil shale deposits hold 10 million Quads; heavy oils are already being extracted

---

separate Nobel Prizes have been awarded for ATP-related discoveries, the most recent, in 1997, for research on the key ATP-synthesizing enzyme; the Nobel Committee compared it to "a water-driven hammer minting coins."

*In 1886, J. P. Lesley, the state geologist of Pennsylvania, declared: "I take this opportunity to express my opinion in the strongest terms, that the amazing exhibition of oil which has characterized the last twenty, and will probably characterize the next ten or twenty years, is nevertheless, not only geologically but historically, a temporary and vanishing phenomenon—one which young men will live to see come to its natural end." Quoted in P. Giddens, *Oil Pioneer of the Middle West* (Standard Oil Company, 1955).

FIGURE 11.2   U.S. Agriculture, Population, and Energy

Source: EIA, *Annual Energy Review 2003*; U.S. Census Bureau, *Historical Statistics of the United States Colonial Times to 1970*; Bureau of Economic Analysis; Louis Johnston and Samuel H. Williamson, "The Annual Real and Nominal GDP for the United States, 1789–Present," Economic History Services, April 2002.

James Watt's coal-fired steam engine was invented to mine more coal. In short order, humanity was harvesting billions of tons of plants that had grown in the Cretaceous and Carboniferous periods some 65 to 360 million years ago. Energy consumption and wealth have since risen exponentially.

by brute thermal force from the Canadian Athabasca deposits, and bio-engineered bacteria could make the Earth's vast deposits of these oils economically accessible everywhere within a decade or less. Even more abundant is the energy locked up within uranium and other radioactive elements. The world's oceans contain over 10 *trillion* Quads' worth of deuterium, a fuel that we will in due course learn to unlock through nuclear fusion.

Nothing very fundamentally new will be required to unlock it. All "production" of energy begins with the scavenging of scattered pockets of high-grade energy from the environment. This is just the first stage in the purification pyramid, the first step in plucking order out of chaos. People who drill oil have to rearrange a mile or two of rock. People who drill for solar power have to rearrange and maintain microscopic layers of silicon, cadmium telluride, copper indium diselenide, and gallium arsenide. Those who drill for deuterium will rearrange cubic kilometers of seawater. Tomorrow, as today, we will begin with diffuse, dispersed forms of energy and end with more concentrated forms. We will consume high-grade energy in the process and extract still more by doing so.

As it always has before, it takes just two things to do that—an intelligently designed excavator, drill, solar cell, plant, or other structure, and energy itself. The structure incorporates logic, intelligence, information, know-how—something or other, which nobody quite knows what to call—that's smart enough to impose order on its energy-rich surroundings. The energy is needed at the outset to power the machine's order-imposing mission.

POWER IN PURSUIT OF ENERGY

Begin with the energy-capturing half of the story. The 1981 *Science* article took the gloomy view: if it takes energy to find energy, you end up with less energy overall at the end of the day, even when the hunt was successful. The sunny view is that the more energy you have in hand at the outset, the faster and more successfully you can hunt. And that is in fact how things have worked out in practice.

James Watt's first mission for his coal-fired steam engine was to mine the coal. Colonel Drake used coal-fired steam to drill for oil. Today's drills are powered by huge truck engines powered by diesel fuel—oil itself. And generators on a mother ship at the surface dispatch electricity at thousands of volts, via monstrous tethers, to the remotely operated vehicles that search for oil miles beneath the surface. Copious amounts of electricity are likewise used to purify the semiconductors—silicon, most importantly—that can bypass the hundred-million-year process of converting sun to oil and capture solar energy directly.

Thus, energy in motion—power—pursues energy that's sitting still. The energy-to-energy connection is self-evident, perhaps, when a pasture feeds the horse that pulls the plow, or when coal is burned to fire the engine that extracts more coal from the mine. It is less obvious when a diesel generator sends electricity down a 5-mile tether to power a laser-radar unit on a remotely operated vehicle. And even less so when a silicon power plant is used to power a super-computer that makes sense of the seismic images that locate the oil. Strip mines and monster trucks are so familiar by now that nothing about them seems very elegant, and the environmental damage left in their wake is not elegant at all. The atomic-scale structures of the fuel cell's PEM and the quantum technologies embedded in photovoltaic semiconductors seem very elegant indeed, and perhaps they will prove kinder to the environment too. Both, in any event, are manifestations of a single astonishing phenomenon: energy in pursuit of more energy, high-grade power being used to build and propel structures that pluck low-grade energy from the environment, in a self-amplifying cycle.

As the passage quoted at the beginning of this chapter suggests, Sadi Carnot himself, father of the second law, drew attention to this fundamental circularity—that the mining of coal was the first use of the coal-fired steam engine. Perhaps—though he did not make this clear—Carnot grasped that this pedestrian commercial fact came perilously close to contradicting the grand scientific law he had just propounded. The second law says that high-grade energy inevitably gets dissipated and dispersed. Yet put to practical use, the steam engine does just the opposite—it uses a bit of coal to get a lot. With the help of the right engine, located on the surface of the right planet, a small pile of fuel contrives to create a bigger one.

LOGIC IN PURSUIT OF POWER

It is not enough, however, to establish that energy can be used to capture more energy. One must also resolve the issue anticipated in the 1981 *Science* article on oil: how much energy in hand does it take to capture the energy in the bush? However many gazelles there may be on the Serengeti plain, the lion will starve if it must burn up two gazelles' worth of calories to chase down one. Will one unit of energy in hand capture more than one unit in the bush, or less? So long as the answer is more, all is well, and energy supplies can spiral up. As soon as the answer is less, the end is inevitable—the lion will starve, the drills will stop turning, and the lights will go out.

Which it is—more or less—depends on two things: what's out there, and how clever we are at retrieving it. What's out there we cannot change, but we can get better at retrieving. The energy-seeking engine— the regulator, valve, piston and cylinder, the core internal logic of the material structure—is every bit as important as what lies scattered about in the environment through which it roams.

The first and most direct way to improve returns from the energy-seeking machine is to use whatever energy it does find to build more machines just like it. That's certainly how nature managed to become what it is today—life simply begot more life. Most agrarian societies have functioned that way too—anything not used for survival was used to increase the number of farmers in the next generation.

In just the same manner, some of the coal mined with Watt's steam engine was used to haul coal itself, to places where it was used to smelt the iron and power the engines that were used to build, among other things, more steam engines, which were then hauled back to the coal mines. In *Coal: A Human History*[4] Barbara Freese recounts how the Englishman George Stephenson opened the first functional railway in 1825, to link the coal town of Darlington with the river town of Stockton. "The difficulty of hauling coal has always been one of its greatest drawbacks as a fuel, but now, through the locomotive, coal could haul itself."[5]

Energy-seeking engines have continued to improve ever since, and are evolving more rapidly today than ever before. Satellite, acoustic imaging systems, and data processing play such a pivotal role in today's search for oil that the modern drilling rig has been aptly described by Jonathan Rauch as a computer with a drill bit attached to one end.[6] Brute force is still needed, but drilling for oil has become a delicate, high-precision process of keyhole surgery, conducted by intelligent electrical actuators and movers working miles away from the closest human operator.

Such ingenuity is of course being put to use to capture "renewable" fuels too. The trick for extracting power efficiently from a variable wind is to continuously adjust the pitch of the rotating blades and to feather them when powerful gusts threaten to tear the structure apart. Thus, the magnificent windmills now being built are dexterous robots that use advanced electromotive technologies to pluck power from the wind as intelligently and adaptively as a hybrid car plucks it from a gasoline engine or a da Vinci surgical system plucks it from a distant power plant.

Solar technologies are advancing even faster. The first cell was invented by Charles Fritts, an obscure contemporary of Edison's, in 1883; the first cell to produce useful power emerged from the Bell Labs only in 1954, with the advent of semiconductors.* The space station's four 400-square-meter solar panels are still single-junction devices that capture energy from only a relatively narrow band of colors in the sunlight, at an overall conversion efficiencies of 10 to 15 percent. But a three-junction, gallium indium phosphide/gallium arsenide/germanium PV solar cell recently achieved a stunning 32 percent conversion efficiency. And the price of PV cells has dropped fifty-fold in the past three decades. Notwithstanding everything we wrote in the preceding chapter, engineers will undoubtedly, in time, find ways to incorporate cheap, high-efficiency semiconductor junctions in roofs, walls, and widely used construction materials.

Less clear, however, is whether any of these renewable-energy technologies will improve much faster than conventional ones, as they must, to catch up. The lasers emerging from the same semiconductor fabs that build solar cells can enrich uranium at least 20 times more efficiently than the gaseous diffusion processes currently used. A 10 kilowatt (kW) chain of semiconductor-diode lasers can pump intense pulses of light

---

*The discovery did, however, spark a great deal of interest in the scientific community, from Werner von Siemens to James Maxwell, and not too long after, from Albert Einstein. Indeed, Einstein's Nobel Prize was awarded "for his services to Theoretical Physics, and especially for his discovery of the law of the photoelectric effect"; the Nobel Committee knew it owed him a Prize, but didn't quite dare mention Einstein's radically new theory of relativity. See John Perlin, "Solar Power: The Slow Revolution," *Invention & Technology*, Summer 2002.

into a dye laser, which directs a meticulously tuned 2 kilowatt beam of orange light at a gaseous stream of uranium hexafluoride to separate the heavier (U-238) from the lighter (U-235) isotopes used to fuel a reactor. In the not-too-distant future, even higher-power diode-lasers arrays, pumping ytterbium-doped strontium-fluorapatite lasers, will compress deuterium or tritium to ignite controlled nuclear fusion. Our two-century march from coal to steam engine to electricity to laser will thus culminate in a nuclear furnace that burns the same fuel, and shines as bright as the sun itself. And then we will invent something else that burns even brighter.

Or, if we prefer to keep on digging, the day is not far off when 6-inch diameter pulsed beams produced by advanced high-power lasers will replace rotary mechanical drills. Bundles of optical fiber will channel the energy down the 5-mile borehole, with lenses at the end to focus the laser light on the rock face. The intense heat will melt the rock, extend the borehole, and then sheath it in solid ceramic, eliminating the need for a steel casing. The power of the photon will thus pursue and retrieve fuel created a hundred million years ago by the power of the sun.

No universal law of nature holds that power and logic will improve faster at the top of the energy pyramid than resources recede at the bottom. But energy-capturing technologies are improving across the board, and faster today than ever before. The logic of the fuel-retrieving machines has advanced much faster than the fuels have retreated—we keep getting closer to the receding horizon. Environmental concerns are a separate matter, important in their own right. But the issue of exhaustion is resolved. Energy supplies are—for all practical purposes—infinite.*

---

*In May 2004, with $40-per-barrel oil generating daily headlines, *Science* published a piece arguing that periodic panics about imminent exhaustion are almost as old as oil production itself, and are invariably followed by new bonanzas of production. "The world is not running out of oil"; there will be "abundant supplies for years to come." Leonardo Maugeri, "Oil: Never Cry Wolf—Why the Petroleum Age is Far from Over," *Science* 304, no. 5674 (21 May 2004): 1114–1115.

MAXWELL'S DEMON

So it was settled—Sadi Carnot had worked out the second law correctly, and no one was going to prove him wrong. More than a century later, Albert Einstein would observe that the young Parisian's law "will never be toppled." Except that it had already been toppled, or so it seemed.

In 1871, James Clerk Maxwell proposed a subtle challenge to the second law.* He imagined a demon that could order atoms and molecules much as life appears to do, but in a way that the second law definitely forbids. The demon sits beside a little trap door that separates two chambers, A and B, each filled with the same gas—krypton, let's say. The demon carefully watches atoms bouncing chaotically about in the krypton on side A. When he sees one moving particularly fast, and headed his way, he opens his gate just in time to let just this one atom fly through to side B; he does the same in reverse to let slower-moving atoms cross over to side A. The temperature of the krypton is defined by the average speed of its atoms—so side B gets hotter, and side A gets colder. This requires no effort at all, because the trap door is as light as can be, and moves without friction. But Carnot's law—the second law—says (in one of its several forms) that heat can't do that. Moving heat up-hill always requires some input of higher-grade energy, like the electricity that runs a refrigerator.

The demon perplexed physicists until well into the 1960s, when theoretical work finally laid his imaginary body to rest. Maxwell had con-

---

* "If we conceive a being whose faculties are so sharpened that he can follow every molecule in its course, such a being, whose attributes are still essentially finite as our own, would be able to do what is at present impossible to us. For we have seen that the molecules in a vessel full of air at uniform temperature are moving with velocities by no means uniform . . . Now let us suppose that such a vessel is divided into two portions, A and B, by a division in which there is a small hole, and that a being, who can see the individual molecules, opens and closes this hole, so as to allow only the swifter molecules to pass from A to B, and only the slower one to pass from B to A. He will thus, without expenditure of work, raise the temperature of B and lower that of A, in contradiction to the second law of thermodynamics." Maxwell, "Theory of Heat," 1871.

ceived of his demon as a pure source of logic—as "a being whose faculties are so sharpened that he can follow every molecule in its course,"
but who acts "without expenditure of work." In 1929 Leo Szilard[7] suggested, on the strength of intuition alone, that the demon's measurement
or thought processes somehow had to require at least as much useful energy as the demon could capture by the operation of his gate. Decades
later, other theoreticians turned that hunch into rigorous theory and established a surprising fact: it is in *forgetting* what he did last time
around that the demonic gatekeeper must actually dissipate at least as
much order as he creates each time he actuates his gate.

To do his job, the demon must look for incoming traffic and then
briefly remember what he sees. And to do that, he must store one "bit"
of information until it's time to open the gate to let the hotter atom
through. Then he must "clear" his memory register to start the process
again. The storing requires some ordered energy which the clearing dissipates—and the second step, the forgetting, invariably dissipates as
much high-grade energy as the demon ends up adding to the universe by
activating the gate at just the right moment. A demonic gatekeeper must
dump some chaos in the *forgetting* phase of his cycle, just as a thermal
engine must dump waste heat to increase its own internal order in the
cooling phase of its cycle.

It is always in the shedding of energy that the thermodynamic accounts are honestly settled. Maxwell, reflecting on demons, had over-
looked what Carnot, reflecting on steam engines, had fully grasped: you
always have to shed energy—*always*—to shed entropy and increase order. Waste is not merely inevitable, it is virtuous. Nothing useful can be
done without it.* Even a demon must play by the rules. Even a demon
must waste.

---

*The theoreticians have at least one more great conundrum to solve in this
arena. A mathematical equivalence has been established between statistical "order," at the atomic scale of things, and conventional heat and temperature, at the
macroscopic scale. This mathematical construct, however, cannot distinguish
between (say) a strand of DNA gibberish and a strand of DNA capable of de-
fining a living organism, which is capable of using power to replicate itself. Yet

ROOM AT THE BOTTOM

The demon in Watt's steam engine was the regulator, the structure that opened and closed the valves in synchrony with the movement of the piston. The logic in the silicon selenium junction of a solar cell is a billion times smaller, finer, and faster. The overarching trend in the engineering of energy-capturing, energy-processing, energy-transforming logic is toward structures that are smaller, faster, more efficient, more complex—millions if not billions of times more so than anything seen before. We are, in short, now building gatekeepers almost as tiny, delicate, quick, and smart as Maxwell imagined. The demon himself will forever elude us, but his atomic-scale gate—*that* we do now know how to build.

Richard Feynman glimpsed some of the possibilities of atomic-scale structures six years before winning his Nobel Prize in a quite different field of physics.* Feynman's December 1959 lecture to the American Physical Society was titled "There's Plenty of Room at the Bottom." "What would happen if we could arrange the atoms one by one the way we want them?" Feynman wondered. "What could we do with layered structures with just the right layers?" We would inevitably end up building tiny electronic elements at scales of "10 to 100 atoms in diameter," he reflected, and they would behave very differently from anything we

---

a strand of DNA that defines functional life is, self-evidently, a form of local order (local to this particular planet) that is fundamentally different from a strand of DNA gibberish.

The bridge between conventional entropy, on the one hand, and statistical "order," on the other, will probably come down to how order in fact exists in the material world, not as a mathematical or statistical construct, but as material structures. The logical arrangement of material structures will always be dissipative, in that it must always overcome the irreducible uncertainty (and thus disorder) inherent in moving and positioning individual atoms. An increase in order lowers entropy (locally) only when power orders atoms in ways that are capable of harnessing more power to create more (local) order.

*Feynman's prize was awarded for his "fundamental work in quantum electro-dynamics, with deep-ploughing consequences."

had ever managed to build before. "I can hardly doubt that when we have some *control* of the arrangement of things on a small scale, we will get an enormously greater range of possible properties that substances can have, and of different things we can do."

The first such structure had already been built ten years earlier, in 1949—it was the transistor, of course, and its inventors would pick up their Nobel Prize in Stockholm just a few years before Feynman delivered his lecture. The p/n junction that does the critical work within the semiconductor is only atoms thick, and its operation exploits the bizarre laws of quantum physics. In the half-century since its invention, quantum engineers have learned to assemble an enormous array of such devices from exotic elements in the Periodic Table—silicon, gallium, germanium, indium, and phosphorous, among others. In their chip fabs, the engineers deposit exquisitely fine, hundred-atom layers of material in near-perfect vacuums—just as Feynman anticipated.

We now buy those devices by the trillions to supply us with digital logic. Atomic-scale *power* elements are now being assembled side by side with the logic, on the same chips—photodetectors, LEDs, solid-state lasers, high-speed radio circuits, temperature sensors, and other structures to project, convert, and sense the flow of power. The logic elements are built with silicon, boron, phosphorus, oxygen, and aluminum; the power components with gallium, arsenic, and other elements from columns III and V of the Periodic Table. These "heterojunction" power-logic devices define something altogether new. They use power not just to *reason* but also to *do*. As we have described throughout this book, such structures are now being harnessed to convert and control raw power—in the grid, the drive train of the silicon car, lights and lasers, dexterous robotic surgical systems, scooters, and windmills.

It may seem incongruous to suggest that atomic-scale engineering will determine the future of high-tension cables that span the country, monstrous furnaces and steam turbines that light them, hundreds of millions of piston engines in our cars, the screaming turbines that propel us through the air, the white-hot tungsten filaments that turn night into day, the architecture of our factories, the productivity of our workers, and the entire future of our gargantuan 100-Quad appetite for raw fuel.

But since Watt's invention of the regulator it has always been thus. The cool order of logic has everything to do with the hot, violent, and chaotic world of energy as we use it. Steam had been around forever. It was Watt's regulator that changed everything.

It is these same technologies that are now being set off to pluck energy out of nothingness, or very close to it—to do almost exactly what Maxwell's demon was supposed to do, but honestly. Expose a silicon-selenium solar cell to daylight, and the p/n junction within it plucks power from (almost) nothing to dispatch electricity down a wire. Place a suitable array of power and logic chips behind it, and the high-efficiency wind turbine plucks power from the wind—almost nothing once again. Place such devices in a remotely operated vehicle, and it will set off in search of oil beneath miles of water and rock—oil so distant, so well hidden, that for practical purposes it too is nothing until such technological demons arrive to transform it into something.

Like the DNA in a stalk of corn or the regulator in Watt's steam engine, the new atomic-scale power-logic structures have no obvious connection with the production of energy in the gargantuan quantities that we actually use. The DNA in a kernel of corn doesn't produce energy—energy is required to synthesize it. Watt's regulator produces neither heat nor motion—it uses motion to direct the flow of heat and consumes some power in the process. The p/n junctions in semiconductors do the same with electrons and photons. They channel, switch, regulate, direct, and convert—and they always use power to do so. But at the end of the day, they all contrive to end up surrounded by more useful energy than they consumed at the outset. After a summer of sun and rain, the stalk of corn delivers far more sugar than started out stored in the kernel.

PERPETUAL MOTION

The great cosmic truth challenged by Maxwell's thought-experiment and reaffirmed in proving it wrong was that *design, intelligence, logic*—something of that sort—can impose order on the ambient chaos, so long as it isn't quite chaos, so long as some bits are hotter and others colder, or some chemically richer and others poorer, or some faster or slower, or higher and lower. Maxwell's only mistake was to assume that logic was free.

Maxwell was mistaken in imagining that logic might impose order "without [any] expenditure of work," but that, it turns out, is only a detail. The new structures of power don't actually beat the second law of thermodynamics—they dissipate energy as they open and close their molecular and quantum gates, to let the higher-grade power pass through while they leave the lower-grade behind. But they don't dissipate much. Yes, it takes power to build Maxwell's gate and more power to operate it. But on the planet we happen to occupy, that *is* just a detail. Far, far more energy can be found and captured by such demons than is consumed in building and running them. Like the valves in James Watt's steam engine, the transistor and its powerful progeny impose order on the flow of power much faster and more effectively than has ever been possible before. And by imprinting logic on power so much more efficiently, they make nonsense of everything gloomy that we think we know about energy.

Most of this book has been devoted to the real-world incarnations of Maxwell's demon and their policy implications. We have chronicled how logic and energy transform lower-grade energy into higher, in practical engines that turn shafts, drive generators, propel cars, and project intense beams of light, radio waves, and X-rays. We have told the engineer's story of what has been accomplished in the two centuries since a forerunner of Maxwell's demon came to Earth in the person of James Watt.

In the end, one must understand the demon, and the second law against which he does endless battle, to understand everything else. His mission is to produce the purer power that makes everything else possible, but he needs energy to purify energy. Using energy for that purpose, with all the waste it inevitably entails, is the best possible use, because unless you start by doing *exactly that*, you get nowhere at all. And the farther you want to go, the more energy you must expend purifying energy. With the demons in hand today, it takes a full Quad of raw fuel at the bottom of the energy pyramid to deliver one-hundredth of a Quad of digital logic or laser light at the apex. But it is by powering the demon that we power everything that adds order—comfort, convenience, safety, stability, and years—to our lives.

And we will always want more order, and still more, because order is the essence of logic, knowledge, wisdom, and life itself. From the bacterially enriched uranium reactor at Oklo, to the Spindletop oil field, to the heterojunction semiconductor, the trajectory of all life on Earth has been defined by the self-amplifying process of energy in successful pursuit of more energy. The single incontrovertible historical fact about energy on Earth is that energy supplies haven't run down, they have run up.

Life has risen out of the abyss to cover its surface. The slow drift of the amoeba has become the screaming speed of the falcon. Humanity has accelerated from the shuffle of the Neanderthal to the cosmic speed of the asteroid, and we now have on our drawing boards serious designs for unmanned interstellar craft that will move at measurable fractions of the speed of light. For all practical purposes, in this sweet spot that we occupy at just the right distance from the sun, with the black cold of outer space on the far side, perpetual motion is not only possible but probably inevitable. The logic of the planet's thermal cycle made it so. The logic in the proteins and nucleic acids of life made it so. The logic in Watt's steam engine made it so. The logic in the p/n junction will make it so. We can keep on moving, and at ever increasing speed, for as long as the sun continues to shine, and the planet rotates, and the depths of the cosmos stay cold.

THE POWER OF LIFE

The general idea of classical physics is, we progress toward the running down of the universe. . . . What we see here on Earth is just the opposite. . . . Instead of going to heat death, we see successive diversification. And so, in spite of the fact that the second law is probably satisfied, we are not going toward equilibrium, because this stream of energy comes to us finally from the stars, the galaxy, and so on. It ultimately originated in the big bang or whatever—the original presence in the universe.

—ILYA PRIGOGINE (1983)[1]

AFTERWARD, with the wisdom of hindsight, many scientists were surprised they hadn't thought of it themselves. Kary Mullis did, and he would win a Nobel Prize for his single flash of genius. An obscure biochemist in 1983, Mullis toyed with a problem of nucleic acid biochemistry as he drove along a mountain road in northern California one moonlit night. A sudden insight impelled him to pull over and wake his girlfriend, who was asleep in the passenger seat beside him. He had solved a problem of great significance, he told her—he now knew how to replicate DNA, the chemical that encodes life.

Cetus, his employer, gave Mullis a $10,000 bonus for his discovery of the polymerase chain reaction (PCR). In 1991, the company sold the PCR patent for $300 million. Other companies soon developed machines to automate PCR reactions in the laboratory. Today, the worldwide market for the DNA-replicating machines and the tests they make possible is well over $1 billion a year. It doubles in size about every three years.

SUGAR AND THE HELIX

Life is progress from chaos to coherence, from muddle to order. As Robert Wright observes in *Non-Zero: The Logic of Human Destiny*,[2] a "creative thrust" can be discerned in "all of history since the primordial ooze." While the universe as a whole drifts toward chaos, life has managed to reverse that process in at least one small corner of creation. Over billions of years, here on planet Earth, "more and more usable energy was crammed into small organic spaces, and the total amount of order grew, folded into more and more complex forms." These structures embody information—"a structured form of matter or energy whose generic function is to . . . [direct] matter and energy to where they are needed." And these information-processing, information-sharing structures have permitted life to keep building order "on a grander and grander scale."[3]

But how the whole process got started is quite a puzzle. Nobel chemist Ilya Prigogine explored that fundamental scientific conundrum in his 1980 classic, *From Being To Becoming*.[4] The rising coherence of life requires rising consumption of increasingly high-grade energy. To build a logical structure, you must start with ordered power. It takes ordered power to build nucleic acids or semiconductor junctions. The well-ordered power must be there *first*, to build the logical structure. But the logical structure must be there *first*, too—to extract well-ordered power from the poorly ordered environment.

Just how two equally indispensable *firsts* came together simultaneously at the very beginning remains one of the great unsolved mysteries of our existence. The best story Prigogine could come up with was (in brief) that we do see ordered structures like crystals and snowflakes forming here and there in nature; more organized material structures can,

FIGURE 12.1   DNA

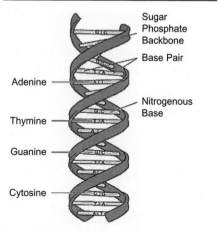

Source: The Why Files, www.whyfiles.org.

Life is a confluence of energy and information.
Stored chemical energy—a sugar, basically—makes
up the backbone of Deoxyribose Nucleic Acid
(DNA). The cross-bars of the helix store information—the genetic code.

in theory, form by pure accident, and if the accident is lucky enough, and suitable supplies of energy happen to be at hand too, in a congenial place, life happens.[5]

It doesn't start with snowflakes however—to oversimplify only a bit, it starts with sugar. In the form we know it today, life is mainly defined by a huge molecule called deoxyribose nucleic acid. DNA consists of two complementary strands, twisted together like long necklaces, each consisting of a complex pattern of beads of four different colors, and twisted together to form a double helix. Each necklace contains the same information, albeit in mirror-image codings. And the backbone of each strand is (basically) a very long necklace of a ribose—a form of sugar.*

---

*A phosphorous-based molecule is also used to help chain things together.

Sugar, as anyone who counts calories knows, is rich in energy, so high-grade energy is required to manufacture it. The plants and animals that DNA defines know how to capture that energy. But what supplied the energy needed to build the first viable strand of DNA itself?

This thermodynamic aspect of the origin of life remains much more perplexing than the strictly mechanical one. The mechanics of DNA replication have been clear since 1953, when the molecule's structure was finally revealed.* Start with two complete strands of DNA, coiled up together as a matched pair. Unzip them. Add the right raw materials, and assemble on each separate strand its mirror image. Then rezip each new pair. One twisted pair produces two. Two produce four, and four eight, and in short order an entire planet can be painted green.

But the process requires high-grade energy to unzip the twisted pair, so as to leave each separate strand structurally and energetically poised to glom on to raw materials in its surroundings, and snap them together into a mirror-image copy of itself. The energy must be supplied very gently and precisely, because the molecules of life are as fragile as they are huge, and because each new copy has to be atomically perfect to preserve the code.

The very earliest forms of life would have been far simpler than any we know today. Indeed, origin-of-life theories do not postulate anything as complex as DNA itself at the outset. All of them do, however, postulate the serendipitous formation of a single structure able both to capture energy and to replicate itself. These models thus assume that two distinct forms of *order* had to converge simultaneously: a proto-biological *structure*, something like DNA, to choreograph self-replication; and *ordered chemical energy*, something like amino acids, sugar, or energetic sulfur

---

*When Watson and Crick finally managed to draw the right picture of the DNA molecule, they immediately grasped how it explains life's fecundity. The self-replicating potential of the double helix was indeed so obvious that they let it speak for itself, making only a single, memorably succinct reference to it in their paper's concluding sentence. James Watson and Francis Crick, "Molecular Structure of Nucleic Acids: A Structure for Deoxyribose Nucleic Acid," *Nature* 171, no. 4356 (25 April 1953): 737–738.

compounds, to propel that same self-replicating process. Additional raw materials (like water and minerals) were undoubtedly essential too, but they were not expected to bring any high-level *order* to the table at the outset, and can thus be more easily taken for granted.

All of the existing theories therefore assume that some energy-rich nutrient soup got formed first. Self-replicating structures then somehow coalesced within it. Under all these theories, one might say, life started out and got a foothold in much the same way as prions survive and multiply today. Biologists don't even yet agree whether a "proteinaceous infectious particle" can even properly be called "life"—the prion is indeed so very strange that for many years mainstream biologists simply could not believe in its existence at all.* They do now. The nutrient soup that feeds the bovine spongiform prion is a cow's brain—this particular prion causes "mad cow" disease. A protein itself, the prion somehow imprints itself on a quite similar bovine protein and, in doing so, transforms it into a second prion. The process escalates until the cow's brain is riddled with holes and the animal dies. All the prions then disappear with it—unless some other animal eats the brain and transfers a prion into a new, protein-rich cranium. The cow's protein is there at the outset. The prion just reshapes it in its own chemical image. The first forms of life contrived to do much the same, in the nutrient-rich soup in which they formed.

But 4 billion years ago there were no cows—so all the existing origin-of-life theories therefore postulate some other mechanism for delivering the first soup. In 1938, Aleksandr Oparin suggested that the carbon-containing compounds that served as the basis of life might have been formed by lightning discharges in the young Earth's atmosphere, which was rich in methane, ammonia, hydrogen, and water vapor. Today, every

---

* When Stanley B. Prusiner first set out his prion theory seventeen years ago, it was greeted with much skepticism precisely because his prions lacked nucleic acids, which were considered biochemically essential for anything "infectious" and therefore "alive." Stanley B. Prusiner, "Prions," *Scientific American* 251, no. 4 (October 1984). In 1997, the discovery won Prusiner the Nobel Prize in medicine.

biology textbook describes the 1953 laboratory tests by Stanley Miller and Harold Urey that lent some credence to that hypothesis. A more recent model looks to radioactivity in the Earth's core as the prime energetic mover of life, by way of thermal hot springs.[6] Yet another model suggests that essential supplies of chemical energy might have arrived on Earth aboard the trillions of tons of meteorites that bombarded the planet's early surface. The meteorites would have picked up their supplies of proto-biological chemicals in the vicinity of the sun or more distant stars, with stellar energy deep in the cosmos thus providing the initial energetic boost to get life out of the starting gate.[7]

In short, all existing origin-of-life models assume that supplies of high-grade chemical energy were synthesized first, indelicately and violently, and then transported to a gentler environment—a warm pond, or a clay substrate, perhaps—where they somehow combined to form organisms capable of self-replication. But the problem with this picture is that while violent high-temperature processes certainly can spawn all sorts of energy-rich chemical by-products, such processes are extremely hostile to anything we think of as life—they rip molecules apart even more readily than they smash them together.

And the longer the postulated pathway separating the first soup factory from the first diner, the more improbable it seems that two could ever have joined forces at all. Particularly when one considers, in addition, that the pathway had to be maintained for as long as it might take for the first self-replicating chemicals to evolve to the point where they could play a more active role in capturing the energy they needed to replicate— quite possibly millions of years. The genius of life is that it unites energy capture and self-replication, in a single, stable, portable, and sublimely gentle package. And it all happens in one place. When all is said and done, the existing models just don't have the look-and-feel of "life" at all.

IN THE HEAT OF THE NIGHT

One of the authors of this book has had the temerity to propose an alternative origin-of-life model: a single, cyclic, chemical reaction that unites the two core attributes of life—self-replication and energy capture—in a single, altogether gentle reaction, propelled by the most ubiquitous source of available energy on the surface of our planet, the mild day-night cycling of the ambient temperature.[8] The indispensable addition of high-grade energy to the replication process could well have been supplied by nothing more than this periodic warming and cooling. No independent source of energy-rich chemicals was needed at the outset; energy-poor raw materials sufficed. So far as the basic thermodynamic requirements go, a hot-cold cycle *alone* is quite sufficient to drive self-replicating chemistry that transforms chaos into order.

Any family of chemicals that folds/unfolds or zips/unzips in response to modest changes in temperature might serve as the linchpin of a diurnal energy-capturing cycle of this sort. And as it happens, nucleic acids do just that, and so do all proteins—they zip/unzip or fold/unfold when the ambient temperature rises and falls. Kary Mullis's flash of genius in 1983 centered on that very fact. In its simplest version, the PCR is a three-minute cycle—a hot stage at 94 to 96°C, which unzips DNA into single strands, and a cold stage at 50 to 65°C in which they rezip. If the right raw materials are present in the broth the rezipping becomes replication instead, and each complete cycle can double the amount of DNA in the reaction vessel.

In the standard laboratory implementation of the PCR, energy-rich building blocks are required as well, along with the thermal cycling, but in strictly thermodynamic terms, they are not essential—*all* the energy required to propel the self-replicative process could be provided by the thermal cycling alone. Raw materials would still be needed, of course—but only as building blocks, not as fuels. The overall thermodynamic accounts would be satisfied so long as more thermal energy is put into the unzipping, at the higher temperature, than is released during the rezipping, at the lower. And that is certainly possible because the unzipping unzips a complete DNA strand, whereas the rezipping simultaneously zips and snaps together the initially separate building blocks.

To exploit the chain reaction to the fullest in the laboratory, you automate the whole process in a PCR thermal cycler. It is a steam engine, one might say, just like Watt's, but this one uses a precisely controlled hot-cold cycle—albeit with an internal-combustion chemical fuel as well—to spin a double helix rather than a steel shaft. Cetus Instrument Systems' first automated PCR machine was dubbed "Mr. Cycle." Today's cyclers are microprocessor-controlled desktop units. Perhaps, then, it was just a bigger, slower, and far more persistent thermal cycler that got everything started 4 billion years ago.

As noted, life today is permeated with biological molecules that reciprocate in response to temperature swings no larger than those that occur every day in shallow ponds across the surface of the earth. Perhaps equally significantly, two-temperature biochemical cycles play fundamental roles in the internal chemistry of everything from single-celled algae to humans. Everywhere that scientists have looked, they have found a clock gene, and one or more complementary clock proteins that adjust and maintain a two-temperature circadian cycle. The circadian rhythm of the fruit fly, for example, has been linked to the delicate affinity of two proteins, two "clock" genes, and a robust 25°C/20°C thermal cycle. Even mammals, which could hold their body temperatures almost perfectly constant, don't in fact do so. The healthy human body, for example, cycles its own temperature between 36.4°C, at around 6 A.M., and 36.9°C, about twelve hours later. Sleep is tightly tied to this thermal cycle, and if it is persistently interrupted (by forced wakefulness), the insomniac soon dies. The most important sleeping apparently occurs toward the end of the cycle, when the temperature is the lowest; eight hours of short cat-naps spread around the clock won't keep you alive.

That modest two-temperature thermal cycles are so inextricably linked to the biochemistry of life today does not, of course, establish that such thermal cycling could ever have provided the energy required for the self-replication of proto-biological molecules. The standard assumption is in fact the opposite—that circadian rhythms evolved later, to adapt living things to an environment that changes a lot between day and night. But to a mechanical engineer, at least, the structures at hand suggest the opposite sequence—structures forming not to *submit* to a thermal cycle but to *exploit* it. Asked to imagine what he might do with molecules that stretch out and coil up, fold and unfold, zip and unzip, in response to a temperature cycle—assuming, of course, that he could work with them directly—James Watt would certainly have set about incorporating them in a new and improved "fire-engine."

Whatever happened 4 billion years ago, we certainly do know that pure "thermosynthesis" of life is possible today. The PCR cyclers that Cetus operates are all powered by the boiler and condenser behind a steam engine 50 miles away; so too are the chemical laboratories that synthesize the energy-rich chemicals that Cetus adds to fortify its reactor soup. The chemicals of life can indeed be manufactured without lightning or meteorites bearing energetic gifts from distant stars. Two temperatures are enough.

Enough for everything. Everything that matters about energy comes down to hot-cold cycles, and structures that pluck their order from the hot-cold interface. Megawatt power supplies feed microwatt junctions to propel logic through semiconductors. Hot feeds into cold to propel motion in steam engines. And the white heat of the sun pours out into the black cold of deep space to propel life on a tiny jewel of a planet that spins on its axis at just the right point between the inferno and oblivion. Perhaps these seeming antitheses are only manifestations of a single, higher logic. Perhaps logic and power are really one and the same. Perhaps they appear different only because we arbitrarily divide our own, human-centered conceptions of existence between heavy and light, body and mind, flesh and spirit. At the time of creation there was infinite power in zero-space and thus, perhaps, infinite logic in that one place and time.

We ourselves, in our own tiny corner of eternity, now strive ceaselessly to recapture fragments of that once infinite logic, and along with all forms of life, we plainly have within us some small power to grasp and replicate it. Nature certainly worked out how to build life from two temperatures long before Kary Mullis did. On a moonlit drive with his girlfriend, Mullis had a Nobel-caliber epiphany that tells us how to set in motion in a laboratory what any man and woman can set in motion, in solitude, on any moonlit night.

United in their solitude, they may perhaps gaze at the sky, and then marvel anew at the astonishing clarity of vision of the man or woman who wandered in the desert over three thousand years ago, and who beheld heaven and earth and surmised that in the beginning the earth was without form, and void, with darkness upon the face of the deep. And God said, let there be light, and there was light, and He divided light from darkness, and day from night, and the evening and the morning were the first day.

NOTES

Preface

1. Paul R. Ehrlich, "An Ecologist's Perspective on Nuclear Power." *Federation of American Scientists Public Interest Report* 28, no. 5–6 (1975): 3–6.

2. Speech to the National Press Club, Washington D.C., August 16, 1979, Congressional Record 9/24/79.

3. EIA, *Annual Energy Review* 2002, Table 7.1: Coal Overview, 1949–2002.

4. V. Badami and N. Chbat, "Home Appliances Get Smart," *IEEE Spectrum*, August 1998; this article provides an excellent engineers' perspective and summary of both historic and future technology-driven gains in appliance efficiency.

5. EIA, *March 2004 Monthly Energy Review*, Table 11.1b: Crude Oil Production: Persian Gulf Nations, Non-OPEC, and World.

6. EIA, *U.S. Crude Oil, Natural Gas, and Natural Gas Liquids Reserves 2002 Annual Report*, Appendix D, Historical Reserves Statistics, Table D1: U.S. Proved Reserves of Crude Oil, 1976–2002; John Fisher, *Energy Crises in Perspective* (Wiley, 1974).

7. DOE/EIA, *Annual Energy Review*, Table 8.2: Counting Solar Photovoltaic Power.

8. *Energy Strategy*, Union of Concerned Scientists, 1980.

9. "Lovins Charts Soft Path," *Electrical Week*, August 2, 1982.

10. "The 'Soft Path' Solution For Hard-Pressed Utilities," Interview with Amory Lovins, *Business Week*, July 23, 1984.

11. *Business Week*, "Germany: The Wilting of the Greens?" August 31, 1998.

12. *Energy: Global Prospects 1985–2000, Workshop on Alternative Energy Strategies* (McGraw-Hill, 1977); *Goals for Mankind: A Report to the Club of Rome* (Universe Books, 1974); John Fisher, *Energy Crises in Perspective* (Wiley, 1974); Philip Hill, *Power Generation Resources, Hazards, Technology, and Costs* (MIT, 1977); W. Clarke, *Energy for Survival: The Alternative to Extinction*

(Anchor/Doubleday, 1974); Yergen Stobaugh, ed., *Energy Future, Report of the Energy Project of the Harvard Business School* (Random House, 1979); Wolf Hafele et al., *Energy in a Finite World: A Global Systems Analysis (Ballinger, 1981); Energy in America's Future: The Choices Before Us* (Resources for the Future, 1979); *Energy in Transition, 1985–2010* (National Academy of Sciences, 1979); H. Landsberg ed., *Energy: The Next Twenty Years* (Ballinger, 1979); John Helm, ed., *Energy Production, Consumption and Consequences* (National Academy Press, 1990); Vaclav Smil, *General Energetics: Energy in the Biosphere and Civilization* (Wiley, 1991); Ruth Howe and Anthony Fainberg, eds., *The Energy Sourcebook: A Guide to Technology, Resources, and Policy* (American Institute of Physics, 1991); Jack J. Kraushaar and Robert A. Ristinen, *Energy and Problems of a Technical Society* (Wiley, 1993); Matthias Ruth, *Integrating Economics, Ecology and Thermodynamics* (Kluwer Academic, 1993); Yoichi Kaya and Keiichi Yokobori, eds., *Environment, Energy, and Economy* (United Nations University Press, 1998); Robert A. Ristinen and Jack J. Kraushaar, *Energy and the Environment* (Wiley, 1998); Robert G. Watts, ed., *Innovative Energy Strategies for CO^2 Stabilization* (Cambridge University Press, 2002).

13. Richard P. Feynman, "What is Science," *The Physics Teacher* 7, no. 6 (1968): 313–320. Feynman explores the same theme at somewhat greater length in *Surely You're Joking, Mr. Feynman!* (Norton, 1985).

CHAPTER 1

1. James Boswell, *Life of Samuel Johnson, LL. D.* (Oxford University Press, 1953), p. 704.

2. Vaclav Smil, *Energies: An Illustrated Guide to the Biosphere and Civilization* (MIT, 1999), p. xv, Table 3.

3. Philip Morrison, "Wonders: Where Fiction Became Ancient Fact," *Scientific American* 278, no. 6 (June 1998).

4. For excellent sources of information on energy fluxes and phenomena, see Vaclav Smil, *Energies: An Illustrated Guide to the Biosphere and Civilization* (MIT, 1999). Additional numbers are set out in Bjorn Lomborg, *The Skeptical Environmentalist* (Cambridge University Press, 1998).

5. Forestinformation.com, "North America's forests at a glance," www.forestinformation.com/beta/Forest_Statistics.asp; Douglas W. MacCleery, "What on Earth have We Done To Our Forests? A Brief Overview on the Condition and Trends of U.S. Forests," USDA/Forest Service, 1994.

6. Vaclav Smil, "On Energy and Land," *American Scientist*, Jan–Feb 1984.

7. For raw fuel prices; Energy Information Administration, Department of Energy *Monthly Energy Review, July 2004*, Tables 9.1, 9.10, 9.11; for capital

equipment purchases, numerous sources including: Annual Survey of Manufactures 2000, U.S. Census Bureau, www.census.gov/prod/www/abs/industry.html; miscellaneous industry sources including 10-K filings of relevant companies (e.g., Agilent Technologies, National Instruments, GE), industry market reports (e.g., Strategies Unlimited, su.pennnet.com; Frost & Sullivan, www.it.frost.com; Darnell Group, www.darnell.com).

8. When we discuss retail prices of electricity and gasoline, we include federal, state, and local taxes.

9. There are a few, very notable, exceptions to the dearth of serious publications on the technology/economic characteristics of electrification in recent decades. See *Electricity in Economic Growth* (National Academy Press, 1986); Philip S. Schmidt, *Electricity and Industrial Productivity: A Technical and Economic Perspective* (Pergamon, 1984); and Sam H. Schurr et al., *Electricity in the American Economy: Agent of Technological Progress* (Greenwood, 1990).

10. A fuller and very readable version of it is set out in Vijay V. Vaitheeswaran, *Power to the People* (Farrar Straus & Giroux, 2003).

11. *Surely You're Joking, Mr. Feynman!* (Norton, 1985) pp. 297–298; the final paragraph quoted here is drawn from Feynman's retelling of the same story in Richard P. Feynman, "What is Science," *The Physics Teacher* 7, no. 6 (1968): 313–320.

CHAPTER 2

1. As quoted in Andrew Carnegie, *James Watt* (Doubleday, Page & Company, 1905), reproduced at the Steam Engine Library of the University of Rochester History Department, www.history.rochester.edu/steam/carnegie/ch5.html.

2. Thomas J. Bergin, "Charles Babbage Lecture," Computer History Museum at the American University. www.computinghistorymuseum.org/teaching/lectures/htmllectures/overview/History_files/frame.htm#slide0032.htm.

3. A highly readable history of the interplay of technology and power appears in R. A. Buchanan, *The Power of the Machine: The Impact of Technology from 1700 to the Present Day"* (Viking, 1992).

4. For a seminal study of the subject, see National Research Council, *Electricity in Economic Growth* (National Academy Press, 1986).

5. C.J.D. Roberts, *History of Babbage's Difference Engine No. 1*, www.home.clara.net/mycetes/babbage/histde1.htm; Doron D. Swade, "Redeeming Charles Babbage's Mechanical Computer," *Scientific American* 268, no. 2 (February 1993): 86–90.

6. See *Electricity in Economic Growth* (National Academy Press, 1986).

7. Peter W. Huber and Mark P. Mills, "Dig More Coal—The PCs Are Coming," *Forbes*, May 31, 1999.

8. Hans Thirring, *Energy for Man: From Windmills to Nuclear Power* (Indiana University Press, 1958), p. 158.

9. Brian Hayes, "The Computer and the Dynamo," *American Scientist*, September–October 2001.

10. For an outstanding text on these issues, see Neil Gershenfeld, *The Physics of Information Technology* (Cambridge University Press, 2000).

CHAPTER 3

1. Sadi Carnot, *Reflections on the Motive Power of Heat and on Machines Fitted to Develop that Power* (American Society of Mechanical Engineers, 1824), reproduced at the Steam Engine Library of the University of Rochester History Department, www.history.rochester.edu/steam/carnot/1943/Section2.htm.

2. N. B. Guyol, *Energy Resources of the World,* Department of State Publication 3428 (U.S. Government Printing Office, June 1949), reproduced in Hans Thirring, *Energy For Man: From Windmills to Nuclear Power* (Indiana University Press, 1958), p. 43.

3. Edward R. Tufte, *The Visual Display of Quantitative Information,* (Graphics Press, 1983).

4. Typically, for example, P. Ehrlich, A. Ehrlich, and J. Holdren, *Ecoscience: Population, Resources, Environment* (W. H. Freeman, 1977), or Union of Concerned Scientists, *Energy Strategies: Toward a Solar Future* (Ballinger, 1980).

5. A. Lovins and H. Lovins, "If I Had a Hammer,"in *World Energy Production and Productivity* (Ballinger, 1981), p. 135.

6. Edwin Houston, *Electricity in Every-Day Life* (P. F. Collier & Son, 1905), p. 298.

7. Jill Jonnes, *Empires of Light: Edison, Tesla, Westinghouse, and the Race to Electrify the World* (Random House, 2003).

8. For a highly readable and engaging book on the second law, see Hans Christian von Baeyer, *Maxwell's Demon: Why Warmth Disperses and Time Passes* (Random House, 1999).

CHAPTER 4

1. Robert Silverberg, *Light for the World: Edison and the Power Industry* (D. Van Nostrand, 1967), p. vi; PBS, Transcript of *Edison's Miracle of Light*, 1995, www.pbs.org/wgbh/amex/edison/filmmore/transcript.

2. Nikola Tesla, "A New System of Alternating Current Motors and Transformers," in *The Inventions, Researches and Writings of Nikola Tesla*, ed. Thomas Commerford Martin (The Electrical Engineer, 1894).

3. Doug Bandow, "Electrical Utilities: The Final Deregulatory Frontier," *The Freeman: Ideas on Liberty*, November 1997, www.fee.org/vnews.php?nid=3898.

4. EIA, *Electric Power Annual 2002*, Table 2.3: Existing Capacity by Producer Type, 2002, www.eia.doe.gov/cneaf/electricity/epa/epa_sum.html.

5. California Energy Commission, "California's Major Sources of Energy," www.energy.ca.gov/html/energysources.html.

6. U.S.-Canada Power System Outage Task Force, including experts from U.S. Department of Energy and the Federal Energy Regulatory Commission (FERC), "Interim Report: Causes of the August 14th Blackout in the United States and Canada," November 2003.

7. National Academy of Sciences, *Making the Nation Safer: The Role of Science and Technology in Countering Terrorism* (National Research Council, 2002), p. 181.

CHAPTER 5

1. Reproduced in Ken New, "1914 Inter-State—'Takes a licking and keeps on ticking'" *Car and Parts Magazine*, May 2003, www.carsandparts.com/features/051903/051903.asp.

2. John Kassakian, "Automotive Electrical Systems—The Power Electronics Market of the Future," Laboratory for Electromagnetic and Electronic Systems, MIT, December 1999.

3. Alfons Graf, "Semiconductor Technologies and Switches for New Automotive Electrical Systems," Siemens, European Automotive Congress, Barcelona, July 1999.

CHAPTER 6

1. Quoted in John Winthrop Hammond, *Men and Volts: The Story of General Electric* (Lippincott, 1941).

2. Jonathan G. Koomey, Alan H. Sanstad, and Leslie J. Shown, "Magnetic Fluorescent Ballasts: Market Data, Market Imperfections, and Policy Success," Publication LBL-37702, Lawrence Berkeley Laboratory, Berkeley, Calif., December 1995.

3. See, for example, Leonard Migliore, ed., *Laser Materials Processing*, (Marcel Dekker, 1996); International Conference on Applications of Lasers & Electro-Optics, Laser Materials Processing, October 2000.

4. See Paul Ballonoff, "On the Failure of Market Failure," *Regulation* 22, no. 2 (20): 17.

CHAPTER 7

1. *Energy Sources,* pp. 131–132.
2. For an excellent survey of emerging technologies, see J. Fouker, ed., *Engineering Tomorrow: Today's Technology Experts Envision the Next Century* (IEEE, 2000).

CHAPTER 8

1. Sir David Brewster, *Letters on Natural Magic* (Harper & Bros., 1835), reproduced in Sir David Brewster, *Letters on Natural Magic* (Arment Biological Press, 2002), p. 190.
2. Lewis E. Lehrman, "Energetic America: The Energy Policy the U.S. Needs," *The Weekly Standard,* September 29, 2003.
3. Jean Gimpel, *The Medieval Machine: The Industrial Revolution of the Middle Ages* (Barnes & Noble, 2003), pp. 11, 12, 25.
4. Matthew Josephson, *Edison: A Biography* (McGraw-Hill, 1959), pp. 372, 378, 429, 430.
5. Commission on Engineering and Technical Systems, National Research Council, *Electricity in Economic Growth* (National Academy Press, 1986); Sam H. Schurr et al., *Electricity in the American Economy: Agent of Technological Progress* (Greenwood, 1990).
6. Joseph F. McKenna, "Aye, robots!" *Tooling & Production,* June 2000, www.manufacturingcenter.com/tooling/archives/0600/0600rob.asp.
7. Gimpel, p. 9.

CHAPTER 9

1. As quoted in Richard Rhodes, *Visions of Technology* (Simon & Schuster, 1999), p. 87.
2. Carlo M. Cipolla, *Guns, Sails, and Empires: Technological Innovation and the Early Phases of European Expansion, 1400–1700* (Sunflower University Press, 1985), p. 131.
3. C. Shannon, "The Mathematical Theory of Communication," *Bell System Technical Journal* 27, March 1948, pp. 379–423.

4. For a long time physicists believed the energy requirements could not fall lower than the background level of thermal noise that exists in any physical structure (a quantity directly related to temperature T and expressed as "kT"). A landmark paper published in 1961 began to put this notion on a formal footing. See R. Landauer, "Irreversibility and Heat Generation in the Computing Process," *IBM Journal of Research and Development* 5, (1961): 183–191; idem, "Minimal Energy Requirements in Communications," *Science* 272, no. 5270 (28 June 1996): 1914–1918; C. H. Bennett and R. Landauer, "Fundamental Physical Limits of Computation," *Scientific American* 253, no. 1 (July 1985): 48–56.

5. The Aluminum Association, Inc., "Industry Overview," www.aluminum.org/Content/NavigationMenu/The_Industry/Overview/Overview.htm.

CHAPTER 10

1. John Habberton, "Of Women, Literature, Temperance, Marriage, Etc.," first published in 1893 and reproduced in Dave Walter, ed., *Today Then: America's Best Minds Look 100 Years into the Future on the Occasion of the 1893 World's Columbian Exposition* (American World Geographic Publishing, 1992).

2. EIA, Office of Energy Markets and End Use, *International Energy Annual 2002,* Table H1: World Carbon Dioxide Emissions from the Consumption and Flaring of Fossil Fuels, 1992–2001.

3. Richard A. Houghton and George M. Woodwell, "Global Climatic Change," *Scientific American* 260, no. 4 (April 1989): 36, 38.

4. David Tillman, *Wood as an Energy Resource* (Academic, 1978), p. 8. See also Brooke Hindle, ed., *America's Wooden Age: Aspects of Its Early Technology* (Sleepy Hollow Restorations, 1975); Joseph Walker, *Hopewell Village: The Dynamics of a Nineteenth Century Iron-making Community* (University of Pennsylvania Press, 1966).

5. USDA Forest Service, Table 3: Forest Area in the United States by Region, Subregion, and State, ncrs2.fs.fed.us/4801/fiadb/rpa_tabler/97_GTR_219_XL_Tables.xls; Forestinformation.com, "North America's Forests at a Glance," www.forestinformation.com/beta/Forest_Statistics.asp.

6. Forestinformation.com, "North America's Forests at a Glance," www.forestinformation.com/beta/Forest_Statistics.asp.

7. USDA Forest Service, Table 3: Forest Area in the United States by Region, Subregion, and State, ncrs2.fs.fed.us/4801/fiadb/rpa_tabler/97_GTR_219_XL_Tables.xls.

8. Steve Nix, "Tree Planting Statistics for the United States," *Your Guide to Forestry,* forestry.about.com/cs/treeplanting/a/tree_plt_stats.htm.

9. Douglas MacCleery, "What on Earth Have We Done to Our Forests? A Brief Overview on the Condition and Trends of U.S. Forests," USDA/Forest Service, 1994.

10. Marlow Vesterby and Kenneth S. Krupa, "Major Uses of Land in the United States, 1997," U.S. Department of Agriculture, *Statistical Bulletin no. 973*, Appendix Table 6: Cropland Used for Crops, 48 Contiguous States, 1910–2000, September 2001, www.ers.usda.gov/publications/sb973/sb973.pdf.

11. Richard A. Birdsey, "Carbon Storage and Accumulation in United States Forest Ecosystems," United States Department of Agriculture Forest Service, *General Technical Report W0–59*, August 1992.

12. United States Forest Service, *RPA Assessment of the Forest and Rangeland Situation in the United States, 1993*, June 1994, p. 12.

13. Robert Dorsch as quoted from August 13, 1999 in "EXXONMOBIL Renewable Energy Resolution," 2002, www.shareholderaction.org/exmob_res.cfm.

14. Peter Huber and Mark Mills, "From Carbohydrates to Hydrocarbons," Grenzen ökonomischen Denkens: Auf den Spuren einer dominanten Logik, ed. Hans A. Wüthrich et al. (Gabler Press, 2001), p. 151; Vaclav Smil, "On Energy and Land," *American Scientist*, Jan-Feb 1984; Marlow Vesterby and Kenneth S. Krupa, "Major Uses of Land in the United States, 1997," U.S. Department of Agriculture, *Statistical Bulletin no. 973*, Appendix Table 6: Cropland Used for Crops, 48 Contiguous States, 1910–2000, September 2001; Data are for contiguous U.S. states for the stated categories of buildings and urban areas, roads, harvested cropland, pasture and range and all energy infrastructure; data do not include, for example, "unused" land categories such as desert, swamps, lakes, streams, etc.; the 2 acres/capita does not include forest land (net use of which for timber extraction is zero since U.S. forests have net growth and in any case timber removed from all the forests consumes just 2% of the "growing stock inventory" (biology.usgs.gov/s+t/noframe/m1103.htm)); nor are National Parks (about 0.25 acre/capita) included in the 2 acres/capita.

15. "On Energy and Land," p. 17.

16. U.S. Department of Energy Information Administration, *Annual Energy Review*, 2003, Table 8.2a, www.eia.doe.gov/emeu/aer/electr.html.

CHAPTER 11

1. Sadi Carnot, *Reflections on the Motive Power of Heat and on Machines Fitted to Develop that Power* (American Society of Mechanical Engineers, 1824),

reproduced at the Steam Engine Library of the University of Rochester History Department, www.history.rochester.edu/steam/carnot/1943/Section2.htm.

2. Charles A. S. Hall and Cutler J. Cleveland, "Petroleum Drilling and Production in the United States: Yield per Effort and Net Energy Analysis," *Science* 211, no. 4482 (6 February 1981): 576–579.

3. Robert P. Crease, "What Does Energy Really Mean?" *Physics World*, July 2002.

4. (Perseus, 2003).

5. Ibid., p. 91.

6. Jonathan Rauch, "The New Old Economy: Oil, Computers, and the Reinvention of the Earth," *Atlantic Monthly,* January 2001. See also P. W. Huber and M. P. Mills, "King Faisal and the Tide of Technology," *Forbes Global,* November 16, 1998.

7. L. Szilard, "Uber die Entropieverminderung in einem thermodynamischen System bei Eingri_en intelligenter Wesen," *Z. Physik* 53 (1929): 840–856.

CHAPTER 12

1. Ilya Prigogine, *Wizard of Time, The OMNI Interviews*, ed. Pamela Weintraub (Ticknor & Fields, 1984), pp. 333–349.

2. (Pantheon, 2000).

3. Ibid., pp. 244–245, 250.

4. (W. H. Freeman, 1980), p. 123.

5. Ibid.

6. D. S. Stakes and J. R. O'Neil, "Earth Planet," *Earth and Planetary Science Letters* 57 (1982): 285.

7. For a readable survey of the competing theories, see Christopher Wills and Jeffrey Bada, *The Spark of Life: Darwin and the Primeval Soup* (Perseus, 2000).

8. For a more detailed discussion, see Peter Huber, Thermosynthesis, A Binary, Periodic, Thermo-Chemical Model for Energy Capture, Replication, Error Correction, and Biochemical Infection, www.digitalpowergroup.com/TBW/thermosynthesis.html.

INDEX

209

PETER W. HUBER is a Senior Fellow at the Manhattan Institute's Center for Legal Policy, where he specializes in issues related to technology, science, and law. His previous books include *Hard Green, Liability,* and *Galileo's Revenge.* He lives in Bethesda, MD.

MARK P. MILLS, a founding partner of Digital Power Capital, is a physicist whose career began in integrated circuits and defense electronics, where he holds patents, and later in energy technology policy, including a stint as a staff consultant to The White House Science Office. He lives in Chevy Chase, MD.